COUNTRY DA

Country Dance

written and illustrated by
HENRY BREWIS

Farming Press Books

A catalogue record for this book is available
from the British Library

ISBN 0 85236 244 7

Published by Farming Press Books
Wharfedale Road
Ipswich IP1 4LG
United Kingdom

Distributed in North America
by Diamond Farm Enterprises,
Box 537, Alexandria Bay, NY 13607, USA

Typeset by Galleon Photosetting, Ipswich
Reproduced, printed and bound in Great Britain by
BPCC Hazells Ltd
Member of BPCC Ltd

Introduction

In the lifetime of many older farmers the fortunes of agriculture have changed dramatically at least three times, – from Reject, to Saviour, to Villain. Maybe even a fourth change already, – to Political Embarrassment. They hardly know what to do with us now.

It's not easy for a peasant to get the steps right all the time, – not when the tune keeps changing. You can look pretty silly rockin 'n' rollin' if the band's playing an old-fashioned waltz. You might even do yourself an injury.

This is only a make-believe story of country folk old and new ... about a small family farm and what happened to it through changing times. But it could be true ... probably is. ...

Henry Brewis

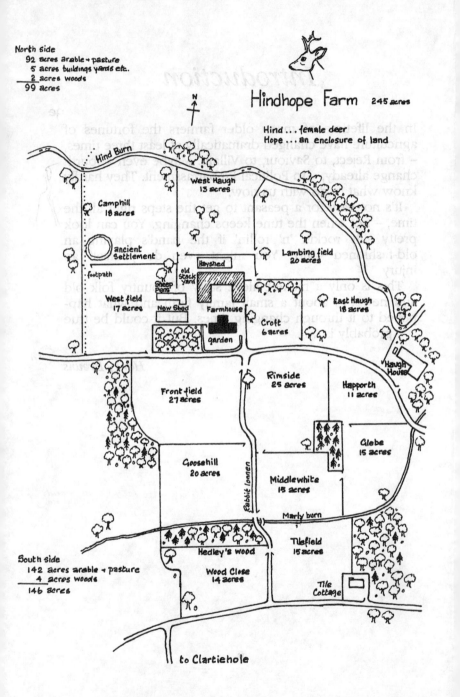

North side
92 acres arable + pasture
5 acres buildings yards etc.
2 acres woods
99 acres

Hindhope Farm 245 acres

Hind ... female deer
Hope ... an enclosure of land

Hind Burn

West Haugh
13 acres

Camphill
18 acres

ancient
Settlement

footpath

Lambing field
20 acres

Hayshed

old
Stack
yard

Sheep
Pens

West field
17 acres

New Shed

East Haugh
18 acres

Farmhouse

Croft
6 acres

garden

Haugh
House

Rimside
25 acres

Happorth
11 acres

Front field
27 acres

Glebe
15 acres

Goosehill
20 acres

Rabbit lonnen

Middlewhite
15 acres

Marly burn

Hedley's wood

Tilefield
15 acres

South side
142 acres arable + pasture
4 acres woods
146 acres

Wood Close
14 acres

Tile
Cottage

to Clartiehole

Hindburn Village

Hind Burn

Marlyburn

Haugh House (Pillick)

Burn Cottage (Forsythe)

Copper's House (Dawson)

Garage (Peter Foggin)

White Hart (Jack 'n' Nora)

West Cottage (Swanson)

the field

School House (Beresford)

The Old Shop (Little)

The Forge (Pratt)

School Master's House (Herbie Flood)

Geordie's Croft

Glebe Cottage (Tommy Cleghorn)

Marlyburn field

Village Farm (Dodds)

Vicarage (Graham)

St. Mary's CHURCH

Sheep pens

War Memorial

Kiosk

Village Hall

Paddock House (Peabody)

The Bungalow

Joe Anderson's Store

'the ranch'

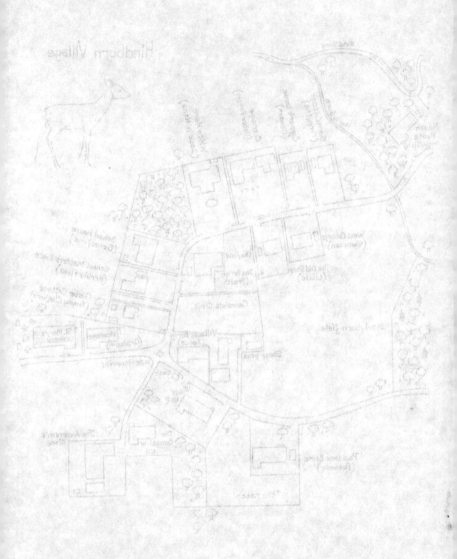

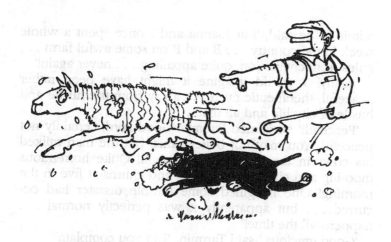

a *Hennessarian*

1

They were drinking in the cocktail bar of the Jolly Farmer just off Waverley Road, – winding down after another tortuous day in the office. Jeremy and Tarquin met there every evening at six, before catching the train home to suburbia.

'Mustn't stay long,' said Tarquin, 'we're out to dinner tonight . . . some friends in the country . . . they've bought an old farmhouse, miles from anywhere.'

'Yes, that's the trouble with the countryside,' Jeremy declared. 'It's so far away . . . another world really, – one might as well be abroad.'

'Worse than that, dear chap, – there's virtually nothing there. I find it all so very boring, a bloody desert. Which reminds me, I think it's your round. . . .'

'Hardly a desert,' said Jeremy, waving his glass at the barmaid.

'I mean it *is* usually green, or yellow or brown . . . things *do* grow there you know, – trees and wheat, stuff like that . . . lots of golf courses. . . . Ever played Bellingham?'

'No,' said Tarquin. 'Tried once, couldn't find it.'

His friend paid for two more large gins. 'I hesitate to

admit it,' he said, 'but Joanna and I once spent a whole week in the country . . . B and B on some awful farm . . . miles from civilisation, quite appalling . . . never again!'

'Really? I would imagine it might have been rather peaceful, therapeutic even . . . away from the hustle and bustle of city life and all that. . . .'

'Peaceful!' exclaimed Jeremy. 'Good Lord, certainly not peaceful. You may not believe this, but the man milked his cows in the middle of the night! Quite horrendous moo-ing and clanking of chains and churns at five in the morning! One imagined some terrible disaster had occurred . . . but apparently it was perfectly normal . . . happens all the time!'

'Good gracious,' said Tarquin. 'Did you complain?'

'Oh of course,' Jeremy said sternly. 'I said to the farmer chappie, "look here," I said, "this really isn't good enough, – we're paying a damn fortune for a spot of rural tranquillity, and all hell breaks loose while it's still dark! . . . Can't you delay this ridiculous operation until after lunch?" ' Jeremy paused to light a small cigar. 'The man was actually rather offensive,' he said; 'told me if I didn't like it I could bugger orf!'

'Charming,' said Tarquin.

'Oh the fellow was totally uncouth,' Jeremy went on. 'His vocabulary was extraordinary, consisted almost entirely of obscenities, – especially when dealing with sheep.'

'Heavens, – there were sheep as well?' Tarquin seemed genuinely surprised.

'My dear fellow there were sheep, cows, hens, dogs, horses, cats, – all sorts of weird creatures, a veritable menagerie. Of course the children were enthralled . . . followed the old farmer all over the place, every day. . . .'

'So at least they had a fun time?'

'Well perhaps,' said Jeremy, 'but I'm afraid the experience may have a long-term psychological effect on them both . . . Clarissa especially has been using some unfortunate words lately. She apparently became far too

2

familiar with the farmer's son – I believe he was called Willie – no more than ten or eleven years old . . . about the same age as Clarissa I suppose . . . but my God the boy was already sex mad! Joanna found them together one afternoon sitting in a hedge, watching some cows behaving quite disgracefully in public . . . Willie explaining every sordid detail!'

'Golly,' said Tarquin, draining his glass. 'I think we need more gin.'

'Young Rupert wasn't spared either.' Jeremy's terrible memories were flooding back. 'Have you seen that programme *One Man and his Dog* on TV?' he asked.

'Lovely programme,' smiled Tarquin, 'amazing relationship between man and beast. . . .'

'It's a fraud!' said Jeremy. 'The real thing's nothing like that at all. This farmer fellow, Weatherspoon or Weatherburn I think his name was, he brought sheep into the yard almost every day, – constantly castrating them, or injecting them for some awful disease. He had a dog called Spot, and I tell you the animal was totally deranged . . . a potential killer! And not at all surprising really, – Weatherspoon would scream some command at the poor thing, and it would immediately tear orf to bite any unfortunate lamb silly enough to attempt an escape. . . . The dog went about permanently with a mouthful of wool. It was like some awful horror film, – the Hound of the Weatherspoons. As you know, on telly the shepherd simply blows a whistle and the dog sits down. Not with this man. . . . He swore and shouted incessantly, even threw sticks and stones at the unfortunate beast. . . . It was no surprise to me that the dog was a menace . . . thoroughly disturbed I shouldn't wonder.

'Rupert, however, thought it all very amusing . . . now he tells our little Corgie to "git away bye" . . . and calls him a "donnert hoond." I've no idea what that means of course, and neither does Rusty, – he's completely bewildered!

'Last Sunday afternoon we went for a little walk in the

park, all the family . . . Rusty goes orf to chase a duck, and Rupert yells "Sit doon y' brainless sod!" – or at least that's what it sounded like. Everyone heard it for miles around, and the poor dog ran away. We only found him on Tuesday, hiding under a rhododendron bush, – still trembling.'

'Gosh, how embarrassing,' said Tarquin.

'Oh I assure you that week was undoubtedly the worst week of my entire life,' said Jeremy. 'The countryside is all very well if you can find a good five-star hotel somewhere, with double glazing, a decent cellar and heated pool. But take my word for it, never stay on a farm . . . !'

'Well I'm astonished,' said Tarquin. 'I imagined it would be so healthy, – a clean, invigorating diversion from stress and urban pollution.'

'Nonsense!' exclaimed Jeremy. 'It was absolutely filthy! I recall we set orf for a little stroll on the first evening, and my God there were animals abluting everywhere, – they have no control at all you know . . . Weatherspoon even collected the stuff . . . had a massive lagoon to store it in . . . and then he spread it all over the fields! Can't imagine why, – because the fields appeared to be knee deep in dung already. . . . We were obliged to tiptoe for miles across the meadow . . . and then a big black and white cow coughed just as Joanna was behind it, – I tell you the beast emptied its bowels in a straight line for fully fifteen yards! Unfortunately Joanna was much too close!'

'Oh lord,' said Tarquin, 'how appalling!'

'Old Weatherspoon thought it a huge joke, larfed like a drain . . . said it was very lucky. Of course *he* was permanently covered in it anyway . . . reeked of the stuff all day. . . .' Jeremy snarled.

'Didn't you get any rest at all then?'

'The children did . . . they were zonked out every night, but Joanna and I never slept a wink all week. At first it was the deafening silence, – we just lay there listening to it. . . . Where was the traffic? The police siren? The drunken revellers being sick? – There was nothing! Then

4

just as we began to nod orf, totally exhausted, some idiot bird began to chirp . . . then a dog would bark, a tractor started up, Mrs Weatherspoon would feed her hens, old Weatherspoon would call his bloody cows in, and bang a drum. . . .'

'Bang a drum?'

'Well he rattled a pail and shouted something, – that was the signal for all hell to break loose, and the world to wake up. *We'd* never been to sleep! Furthermore if we weren't down for breakfast by eight o'clock, everyone had gorn, – vanished to inject more sheep, spread more muck, cut grass, spray barley, or whatever. . . .'

'Oh he grew barley as well, did he?' Tarquin asked.

'Well I suppose so,' said Jeremy. 'To be honest it's all corn to me. You wouldn't know the difference between barley and wheat, would you? It's all just corn . . . acres of it, all riddled with fungi apparently. Old Weatherspoon was complaining about aphids and mildew and footrot, and all kinds of obscure plagues I'd never heard of . . . but then he complained most of the time about something. . . .'

'I expect he was making an absolute fortune in subsidies,' said Tarquin cynically.

'Oh I imagine so,' Jeremy agreed. 'He certainly lived very well, ate enormous meals, prodigious appetite. . . . He went orf to market in a vintage Land Rover, probably worth a mint!'

'Was he green?'

'What on earth do you mean, Tarquin?'

'Well, was he environmentally aware?'

'Good God I shouldn't think so. . . . I told you he spread slurry all over his land, injected anything that moved, – and sprayed the rest!'

'Heavens, – it sounds positively terminal. . . . I expect you were rather relieved to get home to normal surroundings and normal people?'

'Oh we *were*, old chap, – quite delighted to have survived the experience, actually. Take my word for it, farming is another world, a total mystery to normal civilised people like you and I. The countryside's all very well to motor through on a Sunday afternoon, but for God's sake let's leave farming to the farmers I say. . . . In my experience they're all mad anyway. . . . In fact it's probably an essential prerequisite for the job . . . has to be on the top of any peasant's CV, I imagine. . . .'

'Must go,' said Tarquin. 'Dinner at eight – if we can find the place, of course . . . somewhere called Hindhope . . . not even on the bloody map. We'll probably get lorst . . . may never be seen again.'

'Oh I expect you'll be at your desk on Monday morning,' smiled Jeremy. 'Just don't be persuaded to spend all night out in the bush, that's all. It's hell I tell you!'

2

The date 1786 is still there, carved into the stone lintel above the front door. That was the year Hindhope Farm was 'christened'.

There was a rural revolution going on at that time . . . one of farming's flourishing periods, – more and more land enclosed, and new cultivation methods developed. Charles (Turnip) Townshend had perfected a four-course rotation in Norfolk. Down at Prosperous Farm in Berkshire, Jethro Tull invented his seed drill. By 1800 the country was carrying nine million people . . . and nine million sheep. Ten years later Mr Colling sold a Shorthorn bull for a thousand guineas to four gentlemen from Darlington. Agriculture was a fashionable science . . . and deer were drinking from the Hind burn.

The land was there long before that of course . . . plundered, ploughed, courted and jilted by Barons and Barbarians. The county archives list assorted gentry and the church as owners over the years. But there was a time (not so long ago) when nobody wanted to farm it.

In the great depression between the wars Hindhope just lay there for a year or two unhusbanded, widowed in weeds. Hawthorn hedges wandered aimlessly, drains choked and bubbled, and fencing posts, rotting at the roots, fell like wasted soldiers.

Old Norman Harbottle remembers hunting over the farm as a lad. There was certainly plenty of cover for foxes then, he reckons. Whin bushes, bracken and brambles had begun to creep out from the woods, maybe twenty yards into the fields. The boar thistles were so strong, he claims, chaffinches were nesting in them, and when they seeded, down floated on the breeze over half the parish like a gentle July snowstorm. Norman would have been travelling at speed, sixteen hands above ground level, but he still clearly remembers 'a ghost farm' . . . a sort of rural Marie Céleste, – anchored and abandoned on a sea of grey-green fescues, dogstail and Yorkshire fog.

Man had moved out and other creatures moved in. In no time at all (he'll tell you) the farm was home to regiments of rabbits, squadrons of flying pheasants and all sorts of wilderness wild life. A local youth with a ferret, a net and a lurcher could easily fill his mother's larder for nowt.

The barns and byres were silent then, sheltering only rats, spuggies and spiders . . . and the old farmhouse never said a word either, just stared south, – waiting for another peasant to gamble again.

Malcolm Forsythe took the risk at the May term, 1929. The landlord asked no rent for the first year. 'Just farm it,' he said, 'and we'll see how it goes . . . things are bound to get better.' It was an agreement without a contract, a partnership without promises . . . and it was Malcolm's chance to see if this rung on the ladder would hold his weight.

Some might argue it was not a very promising time to start out on your own. Others would say, when everything's on the bottom, there's only one way to go. It was still a gamble.

His first steps had been steady enough. Seventeen

years on the same farm just north of the border . . . to begin with as 'the odd laddie', later graduating to full-time shepherd. The apprenticeship meant working with the odd horse, and doing most of the 'shitty jobs' . . . but shepherding (if you were any use) had prospects and rewards.

And he *was* good at it. A pure natural livestock man, a 'good kenner', with a sharp eye and a smart dog. The wages were never extravagant, but neither was the life-style . . . and the bonus for a canny lambin' was a few yowes of his own. (His wee flock always reared twins of course.) The reward for a successful calving could be a couple of bull calves, 'packs' that invariably topped the back-end sale. It wasn't long before Malcolm was obliged to take a grass park, – and a wife.

When he and Agnes eventually moved south they had already accumulated a hundred ewes in regular ages, twenty-two blue-greys with calves at foot, a collie bitch with five pups, and two growing lads, – Samuel and Thomas. And they hadn't borrowed a penny piece.

Malcolm had been looking for a place to get started for a year or more when he got wind of Hindhope from a pal at Hawick market. There were quite a few farms available in those lean years, but some were too big for him to handle, others no more than smallholdings likely to break your heart. This one sounded about right, – and the terms were certainly attractive.

He went to see it five times before shaking hands on the deal, – walking back and forth through every field, turning over the turf with a spade, finding where clay or rock began, running soil through his fingers. He knew full well it would never be the 'land o' Goshen', but a useful farm nevertheless. Given a bit of luck, he reckoned they could make a fair livin' there, – and that would be good enough. 'Enough and answerin' t' neabody'.

The farm rises to about six hundred feet at the highest point on Camp Hill, and straddles an east–west ridge. Roughly 145 acres falling away to the south, a hundred sloping north to the burn.

It could be 'a cad place', some locals said, a bit on the wet side, and on moderate winter days thin winds can still blow whatever's on the menu in from the North Sea. Hindhope certainly never broke any records for cereals, and the last field of oats was often cut well after the Harvest Festival. It was one of those farms where knowledgeable pigeons and crows often assembled as seedlings emerged, where rabbits might resist 'the myxie' . . . where dark clouds seemed to linger at haytime. And yet, they would tell you, lambs and store cattle shifted well, and grew on to make a bob or two for somebody.

The big plus for Hindhope Farm was always her geography. The house and steading sitting more or less in the centre, looking down Rabbit Lonnen, to the bottom road and Tile Cottage, where the clay workers once fashioned tiles and field drain pipes. Annie Jenkins lives there now, – sometimes alone. A big cheerful widow-woman, her various needs satisfied by a succession of male lodgers, and some domestic work in the bigger, richer houses round about.

The Lonnen, once no more than a cart-track, is now a narrow tarmac lane (where courting couples and bramble-pickers park) and links the two 'proper' roads to form a rural H, – and a hard, dry route to most fields on the farm.

Up on Camp Hill you might find the scars of earlier settlers. The University's archaeological faculty once spent a whole summer there, scratching about on a two-acre site. They found the remains of a little 'village', eight round houses, each about fifteen feet in diameter, with cobbled floors and signs of a fire in the middle. Students trowelled up broken pots, pieces of jewellery, bones, tools and weapons, – gradually building a picture of the folks who lived on the hill centuries ago. The 'Prof' concluded it was pre-Roman. Sure enough Hadrian's legions had been stationed hereabouts, but the ancient Brits had suffered a few lambin's before they arrived, he said. It was a reminder (if anybody needed it) that Malcolm Forsythe's lease (or anybody else's for that matter) was little more than the wink of an eye.

The Camp Hill field lies north-west of the house, and backs onto the Hind burn, which serves as the northern boundary of the farm and the parish. The stream also divides the hunting Milford estate on the south side from the shooting Nethercote land to the north. Two feudal empires with opposing religions. One worshipping the fox, – because it's fun to dress up and chase him . . . the other viewing the rascal as a terrorist who is partial to young partridge and pheasants.

Three other fields lie along the burn side, rigged and furrowed pasture for as long as medieval memory . . . ideal territory for yowes to lie on their backs. Even ploughing grants failed to persuade Malcolm or the 'War-Ag' to turn them over. They knew the tips of flintstone peeping out here and there were as big as stately homes underneath. And no one had the heart to chop down those big old trees either: ash, sycamore, a few noble beech, standing idly about as if in Ducal parkland. There they'd been for generations of Shorthorns and Herefords to shelter under and scratch themselves, polishing the bark on the low branches. These fields were known as the West and East Haughs (meadows by the river) and the 'Lambin' Field,' a sheltered maternity ward leading from the croft. A place for intensive care, frantic midwifery and bad language on rough March nights.

South of the road, eight other fields of varying sizes stretch down on both sides of Rabbit Lonnen to the low road. They all have names. *Tilefield*, where the clay was dug . . . a wet hole there yet, even in dry summers. In winter it could be 'bottomless', and almost every year an unwary aristocrat would gallop his mount into the mire and slurp to a halt. The sight of a purple colonel, heels and hands going flat out on a stationary gelding, up to his belly in bog, was a regular delight for followers of the hunt.

11

The *Glebe Field*, so called because it once belonged to St Mary's Church. *Wood Close* a fourteen-acre enclosure below Hedley's wood (the wood named after a huntsman who came a cropper there years ago). *Goose Hill* got its name from the gorse bushes that grew along the Marly burn; *Middlewhite* (fifteen acres) was named for the colour of the old seeded grass blowing in the wind long ago. The *Happorth* (Harpeth) a word meaning 'army's path' . . . and sure enough Romans marched this way to Camp Hill and beyond. *Rimside* (sloping land), twenty-five acres leaning south and east; and the big thirty-acre *Front Field*, so named because, well, that's where it was, – in front of the house.

The steading stood over the road at the top of the lonnen. A U-shaped cluster of stone buildings and the corrugated iron hayshed Malcolm built in '48, all huddled around the yard. Up one side, a series of loose-boxes which used to be stables for half a dozen horses. Years after they'd gone some people reckoned they could still smell them. You might have found a collie dog at the far end, in his own 'suite', – the bottom of the door chewed away in fits of lonely frustration . . . almost an escape hole . . . if only his nose could move the four-stone weight put there to keep him at home. The other boxes were used to house a few stirks, – hell to muck out. A place to set a calf onto a reluctant heifer (somewhere to tie her up) . . . and a sickbay leading out to daylight and a rusty cattle crush.

On the other leg of the U was the cartshed with its rounded keyed arches of sandstone, once a parking lot for bogies, later home for the Fergie with its link box umbilically attached. Later still a Massey 35 and various blue Fords sheltered in there. And throughout the Forsythe era you would find a few bags of old fertiliser (solid plastic rocks) some draining rods, a graip, a barrow, pails, twine. If anything was missing, – chances were it would be in that cartshed somewhere. Above, and running the whole length of the building, – the granary, reached by slippery stone steps up which simple supermen once carried sixteen-stone bags of corn, – because it was expected of them.

At the far end was a cattle yard, a hemmel capable of housing perhaps forty wintering beasts. About a third of this building was covered (originally a cow byre), but the rest was open. Through most of the winter the open area became a quagmire, while the dry, covered section filled up with muck to the beams, and cattle lay steaming and chewing the cud. All this stone designed and built when a staff of five or six stout-hearted men forked hay in, and muck out, through low dark doorways, and ate their bait in the stable watched by quiet Clydesdales. When maids milked gentle cows in white-washed byres, – courted there perhaps by fresh-faced farm lads with big hands and small expectations.

The farmhouse had 'character'. At least that's what an estate agent might say. He could go on to talk of panoramic views, brandish such words as 'potential' and 'challenging'. However he might be less eager to draw attention to the little windows, many of which Agnes could never open, – and the dark rooms. A brochure would probably ignore the erratic supply of water, delivered through Victorian pipes, emerging from the taps like brown ale (but with few of the benefits). The old stone walls, though several feet thick, had no damp course, and badly needed pointing. The house was rather like an ageing mistress who, having received few compliments and little encouragement over the years, just stops trying to impress. Not that this lady was ever beautiful, she never had much architectural elegance. Built to serve a specific purpose two hundred years ago, her figure had never changed . . . T-shaped . . . farmhouse-shaped.

Downstairs on the stem of the T was a kitchen as big as a village hall, and often as draughty. Whenever a northeast wind howled in through the porch at the back door, past the smelly wellies and waterproofs, it could lift the lino all the way to the sink. Agnes had replaced the

13

ancient black range soon after the war with a Rayburn, and this was kept alight every day of the year, – often spitting sparks from hawthorn logs onto a battle-weary, cleaky mat. (There was some talk of an Aga in the late fifties, – but only talk.) Still, the Rayburn heated the brown water, cooked the supper, warmed bums, and saved innumerable half-perished lambs who lay before the open oven door until such time as they could suck, stand up and piddle. Sometimes the dog would sneak in behind his master (Agnes maybe too busy to notice) and grab a few precious moments watching the dancing fire, drying his cold wet nose, licking his feet, – until dismissed to the porch, head down, looking as if he'd been sentenced to death. On one occasion the old tom cat crept over a shivering lamb, and fell asleep on the bottom shelf of the oven. Some time later the lamb was taken back to its mother, and the oven door kicked shut. It wasn't until a cake was put in there that Tom could stagger out . . . a little over-done.

As with all farmhouses, this kitchen was the heart. Not only the place where meals were cooked and consumed, but where deals were done, decisions made, visitors received, bills paid, arguments settled, forty winks dozed . . . and where Agnes (and later, Ruby) ruled unopposed.

Back stairs climbed out of the kitchen to a landing and a spare bedroom (once for a servant lass perhaps). In 1952 this was 'modernised' into a bathroom, complete with 'Niagara' toilet, and a massive iron bath with claw feet. From here a few short steps led to three bedrooms at the front of the house, – two with fireplaces ready to comfort anyone laid low. (Ruby was born in one of them.) The front stairs tumbled down to the front door and the two front rooms, – a sitting room and a dining room, both furnished for their purpose, but rarely used. A 'wind tunnel' passage ran in a straight line from here to the back porch. Away from the fire it could be a discouraging house, – and regular visitors, expecting to stay for supper, came suitably dressed for the experience.

14

It was not unknown for a lady to put her coat and scarf on again before beginning an expedition to the bathroom.

However, whatever twentieth-century sophistications it lacked, – Hindhope was home to the Forsythes. The two boys, Samuel and Thomas had been but bairns when they came. Ruby began life there. London, Birmingham, Paris were on other unreachable planets, and held no attractions. Other farmers they knew might grow bigger crops, buy better machines, build asbestos cathedrals for more cattle . . . might even have an Aga and use a butter knife, – live as if they made a profit. Perhaps they did . . . who cared?

Through the 'hungry thirties' the family took no harm. They ate well enough . . . Malcolm killed a pig every year, Agnes milked the cow and made butter. The hens laid eggs, and the pantry was always full of pies, pickles and rhubarb jam. Malcolm got his first suit two years after he took the farm . . . it cost him forty-nine shillings. The same year the first tractor (a Fordson with spade lugs) was bought for £156.

In the world beyond the farm gate Hyperion won the Derby at 6–1 and an Austin twelve de-luxe with leather seats and sunshine roof sold for 225 quid.

*   *   *

To say the war passed Hindhope by is not as far-fetched as it might sound.

Malcolm dug an air-raid shelter in the croft, but no bombs fell within twenty miles, and the hole filled up with water every time it rained. After a bullock fell through the roof, the hole was filled in, and that was the end of it. Somebody in the village had a nephew wounded in North Africa, and the kids pretended to be Spitfires, arms outstretched . . . swooping through the stackyard after make-believe Messerschmitts.

German prisoners came to help with the harvest. They seemed canny enough blokes and worked hard . . . better

15

than the Italians, who seemed very prone to 'back problems', – except when chatting up the women.

Malcolm bought the Fergie (with its one-spanner tool kit), sold super-special heifers to the government, bargained his barley sample at Berwick Corn Exchange, – and even wondered about taking another farm.

3

Those were days when McCormack binders, pulled by paraffin Fordsons, cut the corn and tied up sheaves (most of the time at least) and flicked them out into tidy rows for the lads to stook. Bundles of Plumage Archer, mixed with dry thistles to pierce the hardest hands, and barley awns that crept cross-over into tender belly buttons. There the grain would stand to ripen on for a couple of sabbaths or more, before being loaded onto trailers and carts, and led away to stack.

Stacking was an art form, a skill to compare with welding or masonry, but with no City and Guilds certificate . . . only ridicule if your house of straw should lean too much or (heaven forbid), – cowp! The builder had to compensate for uneven ground and keep the middle full, sheaves angled down to turn the rain. But not too full, you understand, – or you'd never get the damned thing 'topped out'. . . .

'You're a bit heavy o' the low side,' some expert would shout from below. 'We'd better put a prop in here . . . and another on the end . . . just t' be safe. . . .' Up above, the stacker, on knees wrapped in sacking, going round and round, layer after layer, 'tying' them in, – the 'stack header' feeding him the sheaves tossed up from the trailer. Not too fast . . . give the man time to do a good job. . . . After all, his finished works would be on view (and certainly judged) for most of the winter.

On some cold, clear, frosty morning the thresher would arrive. Willie Turnbull's father had the big Ransomes, Sims and Jefferies machine, and hauled it round the byeways from farm to farm. In earlier days a big, black coal-fired steam engine did the job, later a Marshall tractor. He would 'set it up' between two stacks, check the levels, adjust the drive belt, chocks in place, all points manned . . . start 'er up, – and off we go till darkness.

A small army was required. Two men would be forking sheaves from the stack, tossing them with a casual skill to two others on top of the machine (one would be Willie senior, the captain of the ship). They would cut the strings and feed the corn into the drum. Too fast and the machine would complain with an angry 'WHOOOMPH'. There were no safety guards in those days, and it was not entirely unknown for some careless 'feeder' to fall in. The thresher would complain again, – but the victim seldom said anything.

Down below, straw came out in 'bottles' at one end, grain at the other, and the chaff fell from riddles in the middle. Straw was carried away and stacked for bedding, the chaff dragged off on a big sheet to be burned later. Strong men at the corn spouts filling sixteen-stone railway sacks, tying the necks with binder twine, and lifting them onto carts. There was no coffee break until the thresher was moved to another stack.

A dozen or fifteen folk were needed on such a day. Alice Egdell brought her gang of women from Mafeking Gardens; there would be two or three extra men from the village or a neighbouring farm . . . and Agnes gave all of them dinner in the kitchen at noon precisely. . . . Everybody back to work by one o'clock (maybe after a brief

18

game of football in the yard).

At the bottom of the stack, among the flattened sheaves: rats and mice, thousands of them, nests of them. Border terriers and collies having a murderous time. Forks, sticks and boots, – spearing, clubbing and kicking.

'There's a moose run up m' trooser leg!' screams one of Alice's girls, 'help me for god's sake!' Someone like Thomas, noble but naive, would rush forward to save the distressed damsel, grabbing her leg somewhere about the knee as she squealed and jumped about, apparently in terror.

'No, it's gone higher,' she'd shout. 'Higher,' – and the lad would (more tentatively now) grope his way up her thigh in search of the poor, wee, timorous intruder. There was a limit to how far he could go of course, and it was all just a tease anyway . . . part of the threshing day ritual, a kind of apprentice initiation. He would get his face slapped by the 'outraged' damsel, and everybody would laugh. He would feel pretty stupid for a day or two, but next year it would be someone younger, and nobody would warn *him* either.

*　　　*　　　*

Tommy Gorman, Patrick O'Donovan, the Kelly brothers and Seamus Conner brought Irish store cattle over on the ferry every week. Mostly strong Hereford-cross beasts from County Sligo and Fermanagh. They were sold to the local beef farmers at the mart on a Friday afternoon . . . always at a loss, according to the Irishmen. 'Oh Jaisus, oi canna sell at this proice, Mister Thornley . . .' he would plead. But Mr Thornley the auctioneer would knock them down anyway to a farmer who reckoned he'd hung on for too long already. Patrick would hang his head in despair. 'Oi'm a rooined marn,' he would wail, – and plead for his 'starvin' woif and children' . . . but he'd be back the next week.

Tommy Gorman was always immaculately turned out on these occasions, – at least he was before he went into the ring. Like everybody else, he was covered in muck by the time he came out.

Anyway, he always arrived to take his room at the County Hotel the night before the sale, resplendent in newly pressed tweed suit, spotless fawn-coloured raincoat, dazzling white straw trilby, and carrying a cane stick. The hat was his pride and joy, his trademark. Fellow dealers said it brought him luck, and indeed he often seemed to attract the best trade. He would hang it with his coat and stick on a peg just inside the front door of the County, before having his usual nightcap and retiring to bed.

On the morning of one of these weekly sales he rose early, shaved and dressed very carefully as usual and breakfasted on porridge. He was about to set off for the mart to check his consignment of cattle when he discovered the beloved straw hat was missing. There were other trilbys hanging in the hall, and a couple of caps, but these were of no interest to Tommy. His special hat was gone. He was furious, desperate. The hotel was combed room by room. It was nowhere to be found.

By the time he eventually reached the mart the selling had begun, and Patrick O'Donovan was already in the ring. 'Oh begorrah and bejabers,' he was screaming, 'Oi can't possibly sell these great bullocks at this proice . . . it's bloody hoiway robbery!' He was jumping up and down in a very agitated manner, running frantically back and forth to the auctioneer's box to delay the inevitable fall of the hammer. Finally, when it became apparent that Mister Thornley could squeeze no more than fifty-five pounds a head from the tight-fisted assembly, – Patrick tore the hat from his head, hurled it into the shit and sawdust on the floor of the ring and yelled, 'Go on then – give the buggers away if y' must!'

The hat was a spotless white straw trilby . . . or at least it was until the bullocks ran over it as they left.

*   *   *

Malcolm was a useful shot . . . rabbits on the run, partridge on the wing. As the binders cut a twenty-acre field down to a one-acre triangle, and the rabbits were forced out, he would knock a few over with aplomb and no regrets. After all, they had eaten a fair bit of the crop already. A pair of pheasants on the stubble had no chance, and Agnes would hang them in the pantry until the tail feathers dropped out, before she cooked them for Sunday dinner.

Malcolm had the 'shooting rights' on Hindhope, – it was part of the tenancy agreement, a sanction he very much enjoyed.

The old Brigadier's consuming passion was galloping after fleeing foxes. The only things he had ever fired a gun at were Germans, – and that 'game' was over. So Malcolm could walk the length and breadth of the farm, and shoot his supper any night of the week. He liked that privilege, but he almost lost it. . . .

Agnes kept about twenty-five hens in a wooden hut down the croft. Each evening at dusk, when all the Rhode Island Reds had wandered home to roost, she would walk over and lower the slide on the hen house and bar them safely in for the night.

After breakfast one dark February morning she donned duffle coat and wellies as usual, picked up a pail of granary sweepin's, and went to let the birds out again . . . to peck and scratch and wander. On this day the first creature out of the hole was a fox. He'd been in there all night (Agnes must have locked him in). The crafty old dog had killed a dozen hens before the panic-stricken survivors had scrambled out of reach up onto the highest perches, and there they had sat petrified until dawn. The fox, as full as a barrel now, scurried past Agnes, pausing only to give her a wink over his shoulder (or so it seemed) before disappearing through the hedge, and away.

Malcolm was as livid as his wife. The loss of the hens was bad enough but with lambing due to begin in a month, an arrogant old dog fox was the last visitor anybody needed. He carried the gun every day while checking the stock, but never caught sight of the beast. Agnes looked into the hen house every night before she locked the door and lowered the slide.

Malcolm got his chance two weeks later on a crisp Saturday morning while in the middle of his breakfast. The cocky animal was sitting on the garden wall as bold as brass, head up, sniffing the air, looking towards the hen house. . . . Maybe this time he was planning to hide behind the hut, and then nip out to grab the first pathetic pullet to emerge.

'It's him,' said Agnes, 'I'd recognise him anywhere!'

Malcolm quickly loaded the twelve-bore, tiptoed upstairs, where he knew the bedroom window would be already open, and creeping along the floor out of sight below the sill, he gently, slowly, silently lined up his shot, – the barrels only just peeping out of the window.

The explosion was deafening; it seemed to reverberate all round the house. Agnes, who was on the landing, hands on ears, asked, 'Did you get 'im?'

'I did,' said Malcolm. 'He's a gonna!'

The fox had fallen back onto the lawn as dead as a stone. Malcolm would have given him the other barrel if there had been any sign of life, but when they went to look, there was none. 'That might well have saved us a few lambs,' he smiled. 'Now I'll finish m' bacon and eggs. . . .'

'I think you should bury him straight away,' said Agnes. 'Have you forgotten what day it is?'

He looked at her for a moment while he unloaded the gun. 'It's Saturday,' she said, 'and the Hounds are meeting in the village at half past ten. . . . A dead fox lying in our garden might not impress the landlord very much!'

The Hunt moved off at quarter to eleven, after filling up with 'fuel' at the White Hart. They drew the wood south of the village, and immediately ran flat out in a straight line to Hindhope. The hounds yapped and yelped along the garden wall for a second or two, then turned and sped down to the hen house. Malcolm watched nervously from the back yard as they went frantically round and round the hut (Agnes always left the hens locked in when the Hunt met nearby). The bewildered birds squawked and the hounds howled for fully ten minutes.

Eventually the Huntsman called them off and began to move the pack away towards the river. They seemed very reluctant to go.

'I believe you must've had a damned fox after your hens, Forsythe,' roared the Brigadier. 'Don't worry, we'll catch the barstard before the days out, what!'

'Where did you bury that fox?' asked Agnes when peace was restored at last.

Malcolm was sweating. 'Under the hen house,' he said.

4

Samuel found his father lying by the sheep troughs in the lambing field on April 5th, 1953.

No one saw him fall. But at about 11 o'clock Agnes heard him from the back yard while she was bringing logs in, threatening to murder some 'rotten awkward bitch'. Samuel went to look for him at one when he hadn't come in for his dinner. He still had the stick in his hand . . . and the agitated gimmer he had probably been trying to catch was hanging a muckle single nearby.

At the funeral Agnes and Ruby wept, and the rest of the peasant congregation stood about after the service chatting in the spring sunshine . . . apparently reluctant to leave him. They said what a canny fella he'd been . . . a likeable bloke . . . and nobody's fool either. Alfie Sanderson said Malcolm was the most contented man he had ever known, 'never wanted to be anything other than a workin' farmer.' Everybody agreed the marts wouldn't be the same.

Agnes was lost without him. He had been the only man in her life, – the only one she'd ever fancied. They had always been 'matched' (that was her word). They'd fitted well together, – like village hall chairs . . . and he wouldn't go away. The smell of him was still there, the wind that banged the back door brought him into the kitchen. She never moved his wellies from their place in the back porch, and talked to him as she stood alone at the sink.

She followed him three years later. (When she arrived Malcolm probably said, 'where y' been, woman?' She would just smile and put the kettle on.)

At Hindhope the children moved the wellies and got on with life without much apparent fuss or emotion. It was not immediately obvious that Malcolm had left them only half-prepared. They could handle the day-to-day tasks all right; clip a sheep, milk a cow, single turnips, plough a straight furrow, bake a cake . . . whatever. They had been doing those things since they could walk. But the business side, the money, the buying and selling, the bank . . . well now, those were cards Malcolm had always held, – and very close to his chest. Even Agnes never got to see them. Mr MacDonald the accountant said it was a useful hand . . . just enough cash in the bank to pay the rent and any outstanding bills, – and the farm fully stocked. Malcolm had left most of his money running about on four legs.

Samuel was the eldest and he it was who, without much debate, took the reins. 'Nothin' to worry about,' he declared. 'Thomas doesn't like mechanical things, so I'll look after the arable side; he can do the stock, and Ruby keeps house . . . obvious isn't it?' True, it hardly needed spelling out; they all knew how it would work.

Samuel was an uncomplicated man, – big hands, big feet, big open face. It appeared he only had straightforward thoughts, determined by whatever situation he found himself in. A flat tyre or a blown gasket meant a trip to Foggin's garage for repairs. It didn't mean his day was ruined. It certainly was not an occasion for blaming the gods, gnashing teeth, or hurling spanners about, – and then self-consciously looking for them again in a bed of nettles. Sheer bolts could snap, clouds might rain on sweet hay, sows sometimes squashed their litters, sheep died for no apparent reason. He accepted all that, and dealt with it in due course, with a constantly cheerful disposition that defied disaster or any other imposter. Samuel smiled a lot, whistled a lot, and was seemingly immune to aggravation or stress. If he'd been a dog he would have been a labrador. He could drive his brother mad.

Thomas on the other hand was a whippet, lean and anxious. As the lads grew older, the lines on Samuel's face were smiley lines, – on Thomas they drew a worried map. The man was a loner. As happy as he would ever be in the company of cattle, sheep and dogs, – friends who didn't need a conversation to reassure them.

He had even quietly removed the official footpath signpost that stood on the roadside pointing north to Camp Hill. He didn't want ramblers in big boots or bird-watchers with binoculars wandering about his domain.

'I've just noticed that signpost seems to have disappeared,' Samuel said one day.

'Get away?' said Thomas sounding mildly surprised. 'The Council snowplough must've shifted it last winter. . . .'

He swore quite a lot, as if it were part of normal vocabulary, – but then men who lamb mule yowes often do. He mumbled to himself, even in the company of others. 'What was that?' they would ask. 'Oh, nothin',' he would say. Whatever it was, it hadn't been for public consumption anyway. Sometimes he would sit and talk to Moss, just hardly loud enough to be overheard, both of them smiling a little, as if sharing a private joke. Next day he might just as easily lose his temper and throw something at the bewildered animal. On one such day, having already hurled his stick, he threw his cap at the dog and it flew off with the wind. Thomas was obliged to walk over two fields to retrieve it . . . mumbling, hoping nobody was watching.

There had been other dogs in his life of course. Naturally he told Moss about them, but always with gentle reassurances that *she* was the best bitch he'd ever had.

Tess, he remembered, – a canny old collie who was terrified of suckler cows. But then lots of dogs were.

Tweed was useful . . . a bit over-enthusiastic perhaps, especially in his youth. He had been known to rip the lug off a stubborn yow, or hang onto a top lip too long.

Meg, on the other hand, was so slow y' might as well run around the sheep yourself . . . and Thomas often did, – swearing and spitting until he was out of breath.

And there was Jet. Jet was a disaster. Thomas tried for a whole year to train that beast, but made no impression. The dog was keen enough, but utterly brainless. Instead of running out wide to gather, he preferred to take a short cut through the middle of the flock, scattering them to the four corners of the field, most of them upside down and gasping. He would never sit.

In one last desperate attempt to instil some sort of discipline into the crazy hound, Thomas tied together about two hundred yards of baler twine and, having attached one end to the fence and the other in a noose around Jet's neck, set him off up the West Haugh. He watched until he judged the line was about to run out, and yelled 'SIT!' The dog took not a blind bit of notice until the string tightened, flung him around in mid air and nearly hanged him.

After that he wouldn't run at all, presumably concerned he might be garrotted in full flight. Eventually Thomas sold him at the mart after a store cattle sale. 'Just a young dog,' said the auctioneer, 'tremendous potential,' and knocked him down to the only bidder, a young shepherd from the Tow Law area. Thomas did his best to avoid the lad for the rest of his life.

And yet, as different as they were, the two men got on well enough, Samuel blathering incessantly as he helped Thomas to slowly move ewes and lambs across the road from the croft; Thomas silently lifting fertiliser and seed corn onto a trailer, while his brother set the drill, and whistled 'Good Night Irene'.

27

And what about Ruby? Sister Ruby was a happy kid. Of course she could be quiet, a bit like Thomas sometimes, but mostly she was bright and sparkly. Through her carefree childhood she 'worshipped' her big brothers. Father Malcolm had delighted in his young daughter, taking her for walks round the farm of an evening, checking the stock, her hand in his. She remembered his hand was always warm. Agnes maybe just a touch jealous as she watched them from the kitchen window.

By the time Ruby left school Samuel and Thomas were full-grown men, working a full day. She would take their tea to the field at hay and harvest, help Thomas with the lambing, prepare the supper . . . sometimes wishing she was a boy.

It was Willie Anderson from the Store who persuaded her that being a girl had many interesting possibilities, – and after that first evening in the hayshed Ruby quickly developed quite a 'revealing' reputation. From then on she was often seen emerging (severely dishevelled) from an empty loose box, a dyke back, from behind stacks of barley, out of a byre . . . inevitably followed by a grinning callant, astonished at what he had recently discovered. She became known as 'a canny sport'. At sixteen she was plump and bosomy, when the bra design of the age tended to accentuate the interesting areas higher up, and when suspenders could catapult explorers to darker secrets below. At village dances she bounced and giggled her way through adventurous Eightsome Reels, and dangerous Drops of Brandy; she was never short of a partner for the last waltz, or a lift home in the pick-up truck of some bucolic Errol Flynn.

She never married. Nobody asked her.

The brothers remained bachelors too. Cheerful, chattering Samuel nearly took the plunge once or twice, even got as far as a half-hearted proposal. But when Carol Tomlinson burst out laughing he never summoned up enough courage again. That's how it looked from the outside at least. It might have been nearer the truth to say that, in spite of his innocence, he knew when he was well off. For as long as he could remember, Malcolm had made the big decisions, Agnes or Ruby had cooked his meals, washed his socks, changed the sheets on his bed, lit the fire every morning. What else could he want? A game of darts perhaps, a few pints, pub company, a good laugh, a world of big-ends, bearings, and a Massey maintenance manual . . . no worries. By his late twenties he had (unconsciously) abandoned the courting game altogether.

Thomas, however, was attractive to women. Perhaps they imagined he possessed interesting depths. What was he thinking? Was there something exciting, smouldering beyond those still, dark eyes? The shy, quiet man with the long legs, big hands and black curly hair. They fancied him in a nervous sort of way, – never quite sure of him. Nobody ever got close.

Thomas lived in his own world. He was comfortable enough in the company of the family, – just. But a car coming up the road to the farm was a 'threat', and he would hurry off down the field, or hide in the byre, rather than cope with a stranger. Even trips to the mart were uneasy. In the days before traffic made it impossible, he and Moss would drive half a dozen fat cattle five miles along country roads to the grading most Mondays. It was a slow, leisurely stroll, the heifers grazing the verges all the way, the dog ensuring none of them went astray, – only the school bus to disturb the meander into town. Here, the heifers, on very stange territory now, might venture down a back street, into a treasured garden where an irate housewife would shout too loud and wave her broom. Thomas would only mutter something incomprehensible (it might be an apology, no one could be sure) and quickly manoeuvre to the mart, where super-special, special, or A+ would be the order of the day.

The mart gossip, the over-confidence, the noisy whole-sale butchers made him feel vulnerable, and as soon as the cattle were penned he would walk home again across the fields, Moss nosing about for interesting smells, – the two of them happy with each other's company.

This then was Hindhope, – a small, independent 240–acre state, where laws were made in the kitchen, and everyone knew what was expected of them. Mr Macmillan said they had never had it so good.

The men were 'spoilt rotten' by Ruby. Samuel and Thomas had only to sit down at the table, and within seconds a meal was placed before them. Breakfast was at eight, by which time everyone had been up at least an hour, – feeding and checking livestock, preparing some machine, frying bacon, eggs, tomatoes, mushrooms and black pudding. Ruby had tea and toast after the boys had gone out again . . . and sometimes a 'sneaky' cigarette. The two-course dinner at noon would be followed by half an hour of sepulchral silence while they snoozed. Ruby washed up after they went out again. Teatime was at four, – a brief interval for fruit cake and scones, washed down with strong, dark Ringtons tea. At the end of the working day supper was another two-course affair. Cold ham perhaps, shepherd's pie, a stew, followed by a fruit tart and custard, or rice pudding and prunes.

Samuel developed quite a paunch. Always a big lad, he ate everything presented to him, and sat flatulently on a tractor most of the day. Thomas was made by the same firm, but from a different mould. He remained long and lean, quietly burning up the calories as he strode about the fields in heavy black boots. Ruby spent most of her day either preparing meals or clearing away after meals. She kept the old house spotless (as Agnes had taught her), chopped kinlin', reared pet lambs, fed the pig, milked the cow, darned socks, knitted pullovers, patched torn trousers, shopped, paid the bills, possed and mangled the weekly wash every Monday. Occasionally she would look at her brothers dozing by the Rayburn, and mutter without any real malice, '. . . you two don't know you're born. . . .'

*   *   *

For most of their working lives the Forsythe boys (even into their sixties, they were still known as the Forsythe boys) farmed, like everyone else, through decades of almost unlimited demand. The war started it, and successive governments carried on bribing the farmer to produce more and more for guaranteed markets. From Tom Williams to Christopher Soames to Fred Peart and beyond, – the lights were on green. Cattle and sheep were certified and subsidised. There were incentive schemes for growing cereals, grants for rearing calves, eradicating rabbits, building sheds and ploughing old grass. As the years stretched and turned over, Thomas kept a series of little black books in which he scribbled stock numbers, lambing details, assorted facts and figures of their private world.

March 1962. Bought 7 mule ewes 3 crop with twins – £11/17/6.

April. Ran out of hay. Obliged to buy five tons from Sep Robson at £8/–/– a ton. He notes the quality was much too good, and it was all eaten in a week!

July. Ernie Rogerson bales hay at 3½d a bale.

September 25th. 18 gimmers purchased, £12/10/– apiece.

Sold 2 tons 17 cwts 1 qu oats for 18/6d a hundredweight.

31

At Foggin's Garage four gallons of Esso Extra cost £1/1/2. (In January '63 the Bank Rate stands at 4 per cent but Thomas doesn't mention that).

In 1965 he records a Cereal Deficiency payment of three hundred pounds received for about a hundred acres of barley. The grain was sold for twenty-one shillings a hundredweight.

The brothers sell heifers at £8/3/– per cwt. The Fatstock Guarantee Scheme provides a further £9/6/1 for each animal.

In '66 the rent goes up from £2/15/– an acre to £3/5/– and the rates for the farmhouse are £34/18/8. No problem.

Into the seventies, and the mysteries of metrication and VAT. 'Upstairs Downstairs' on the telly. Jim Prior, James Godber, Fred Peart (again), John Silkin, Peter Walker and Michael Jopling stalk the corridors at the Ministry of Agriculture. To the Forsythe boys, no more than faraway London names, newspaper names . . . pale politicians in dark suits who talk a lot.

A three-day week, a power crisis, Britain joins the EEC . . . events which had little impact at Hindhope, where Samuel and Thomas plodded on like a pair of well-fed, blinkered shire horses, following an endless furrow.

Brother Samuel enjoyed being 'corn boss', and the cereal acreage crept up a little each year. Everybody else was doing it. Thomas was not so sure . . . the old barley barons who used to boast two tons an acre were now quoting three . . . there were even rumours of four! Hindhope would never be in that league. But Samuel wouldn't be denied, – buying expensive bullocks was a mug's game, he said, – and so was paying a contracter to cut corn a week late. He bought a secondhand combine, a new seed drill, a set of discs to replace the spring-tined cultivator that pulled all the stones to the surface. And they needed a bigger grain trailer, a couple of augers, a fore-end loading bucket and somewhere to tip all this corn. A big grey shed appeared, – built by the landlord

(10% onto the rent – no tenant right). It would house cattle in winter, there might even be space for ewes and lambs in March. It all made sense, didn't it? Didn't it??

In 1975 the rent rises to £9 an acre. The following year barley is sold for £62 and wheat at £71. ICI No. 10 costs £84.80, and Thomas buys mule ewes with twins for forty pounds.

Feb. '77. Purchase Mazurka seed barley at £155.00 (inc. Royalty). Most of it blew out in August. What was left sold for £81.50 a tonne. Ernie charges eleven quid an acre to combine, and his baling charge is up to 10p a bale. (Samuel's secondhand combine is terminally ill.)

In November '79 the bank rate stands at 17%.

Kilograms and tonnes replace stones and hundredweights, pounds and p's have pushed out shillings and d's. What is a hectare? VAT goes from 8% to 10% to 15%. Inflation tops twenty per cent.

January '82. Sold twenty tonnes Midas barley at £101.

July '83. Twenty lambs graded at 19 kgs make £44 a head. £24 of that comes in subsidy.

In July 1985 Thomas sells 20 kg lambs for £39, of which £10 is subsidy. Barley sells in October for £100.50.

Two years later similar-quality lambs make £34 in the ring, with only £5.20 added in subsidy. . . . Something called a co-responsibility levy is being deducted from grain sales.

By now the second secondhand combine is fatally wounded and parked behind the grain shed, to be picked at by passing scrap men. The barley crops are not good . . . two fields of Joss Cambier wheat have been a disaster. Thomas says they should never have ploughed out the old grass, – Malcolm never did. Samuel argues they're makin' nowt from cattle, and lamb prices are nothing to write home about either.

Thomas's notes have shown no interest in a bomb explosion at Brighton, the rise of Gorbachev or Tina Turner. The brothers display no urgency to see Starlight Express, Godfather II or their accountant. At the White

33

Hart, MacGregor or Gummer take the blame for most things. Mrs Thatcher is everywhere. Nigel Lawson takes the brakes off, and the economy freewheels into a trade deficit of nearly £3 bn. The bank rate is rising again, and the rent goes to £25 an acre!

Thomas was not happy about it, muttered threats directed towards Sir Nigel, the landlord. . . . 'Greedy sod,' he growled.

'You worry too much,' said Samuel. 'Good, auld-fashioned farmers never fail, man, – they go on for ever. . . . It's the clever buggers that go bust. . . . We'll be all right, we've still got livestock runnin' about, the harvest t' come; valuation's goin' up 'n' up. . . . We canna lose, can we?'

It was Mr Thompson at the bank who first suggested it might just be possible.

'It's pressure from head office, you understand,' he said, with what he imagined was a disarming smile. . . . 'But let's face it, your overdraft *has* acquired a stubborn reluctance to come down . . . in fact it seems to rise every year now!'

There was talk of bullocks to sell; 'they're a canny trade,' Samuel assured him. Still a hundred lambs to go, and there was every prospect of a good harvest . . . all that was needed was a few quid to keep the pot boiling till October . . . it was seasonal. There'd never been any problem before!

'Quite,' said Thompson. 'But times change.'

The brothers had never been good at arithmetic, that was Mr MacDonald's job. Surely as long as you had a bit of luck at the lambin', bought decent 'growing' cattle worth the money, got up in the mornin' and did a proper day's work, you'd be all right, – wouldn't y'?

'We don't go boozing,' insisted Samuel. 'We don't eat at posh restaurants, bet on horses, chase foxes 'n' women . . . not like some folk w' know. . . .'

'Be that as it may,' said Thompson patiently, 'but you haven't made a profit for the last two years . . . and not a

34

lot before that. Perhaps we should have had this talk earlier.' He leaned back in his big, black, leather chair, fingertips together, – 'I'm afraid we must insist your borrowings are reduced forthwith!'

'Forthwith' was not a word Samuel was familiar with, but he suspected it meant 'quite soon'. Certainly this predicament seemed to have emerged with indecent haste, – it didn't seem five minutes ago that Thompson was encouraging them to borrow *more*. 'Good business,' he had said, all smiles and handshakes. Increased live-stock numbers, higher yields. . . . 'Of course you can replace the tractor . . . just sign here. . . .' He'd even got the sherry bottle out. Not recently though.

Back home Thomas said nothing, and went off with his dog, looking worried, – but then he always did. Samuel was marginally less chirpy for a day or two, but after a walk through the corn he reckoned the crops were fairly good. Obviously Thompson was just trying to sound important . . . some people were like that.

A week later he called at the garage for petrol, and Peter Foggin asked for cash.

'We've had an account for years,' said Samuel. 'Just put it down like always. . . .'

'Not any more,' said Peter. 'Last week's cheque came back, and I've been gettin' caught too many times lately. . . . It's cash or nothin'.'

That was the beginning. . . .

'How many lambs can we sell?' asked Samuel anxiously.

'Maybe fifty at a pinch,' mumbled Thomas. 'But the price is nowt flash. . . .'

They got forty-seven drawn out. Forty-five of them graded, and two were rejects. The mart withheld a thousand pounds, reminding the lads they still owed for some bullocks bought three months ago, – and the interest was mounting up. Ruby got fifty pounds for the housekeeping.

A few days later they sold ten cattle privately to a dealer called Freddie Tuffnell. He drove a hard bargain.

35

'You'll have no commission to pay, no haulage,' he argued, 'and you'll have my cheque before I drive out the yard. Can't be fairer than that now, can I?'

They accepted about five quid a head less than they thought the beasts were worth, just to keep the money away from the mart company; they also gave Freddie another fiver a head 'luck money', – which they borrowed from Ruby. The cheque went straight into the bank at 9.31 a.m. on Monday morning.

On Wednesday Mr Thompson informed them it was worthless. Freddie was unavailable for comment. Rumour had it he was wanted for some big Common Market beef fraud . . . but nobody could find him.

By the autumn Hindhope was becoming a dispirited place. Stock numbers were down, and so was quality. Very ordinary four-crop warranted mules, instead of gimmers, replaced the old five- and six-crop toothless rejects. A few smaller stirks to rough winter and hopefully finish off the grass were all they could afford. There would be nothing for the Christmas show this year.

The harvest was a calamity. They had cut a few corners with sprays, nitrogen and seed, trying to economise . . . and it rained!

Last year had been the wettest back-end for ten years, – this one was the wettest for eleven. Ernie's combine stuttered through twisted, flat fields of soggy cereals for weeks. The drum blocked up with soil, the knife chewed through rampant cleavers to salvage a thin pickle, sometimes at 25% moisture. The drying charges were horrendous, the weight loss unbelievable, and Agrigrain deducted what the brothers already owed for chemicals and fertiliser . . . plus interest. The final cheque barely paid the cost of putting the winter seed back into the ground again.

And that was not a straightforward task either. It seemed the sun was embarrassed to peep through ever again. Any brief drying winds only hurried in more clouds. It was a struggle to remove the straw from the

36

stubbles, and cultivations were a proper gamble. If you watched the forecast every night chances were you might be persuaded to stay in bed. If on the other hand, you took a risk and ventured out with the discs, you could easily end up with a seed bed like sago.

Two bad harvests on the trot. This year, fifteen boggy acres abandoned altogether. All this on top of a mounting debt . . . and still the bills came in. It seemed the cost of everything they had to buy just went up and up. What they had to sell (at best) stood still.

Somehow by mid November the brothers had seventy acres of barley and thirty acres of wheat plunged in, – all of it patchy, with slugs and crows in attendance. The last of the lambs and a very lean tup were sold in a dull store trade on a rare, dry Friday. Hardly anyone was at the mart, they were all working land. A dealer from the Barnsley area, who referred to everybody as 'young man', bought the entire catalogue, all except a very lean tup, who keeled over and expired below the auctioneer's box, – before he was sold.

Thomas got ten quid from the mart secretary but waited until 'Barnsley' went to the gents, and then ran for the pick-up. He collected Samuel outside the 'Plough' and they sped off home for tea and Ruby's cheese scones.

The scones were on the kitchen table, along with a note from Ruby. It said she had run away with a dip salesman called Wilfred who, now the boys were obliged to think about him, had been a regular visitor for years. He'd always had a silver tongue and a maroon Cavalier.

Ruby wrote she was sad to leave, but she'd had enough of the slave labour. Now that things were hopeless she had decided to elope with Wilfred who had an index-linked pension scheme, a semi in Bishop Auckland and a cocker spaniel. She signed herself 'your devoted sister, Ruby,' – and left an account rendered from the Co-op for £150.

5

The Forsythe brothers were as near broke as makes no matter. Samuel couldn't believe it, wouldn't believe it. Something would turn up. It was almost as if he expected the patron saint of peasants to appear with a handful of magic beans.

Thomas was more realistic, and more morose than ever. He even appeared to contemplate suicide. He went out one very stormy evening with the twelve-bore, saying rather dramatically, 'I may be gone for some time. . . .' But he only fired one barrel at a rabbit, missed, and came back twenty minutes later soaked to his Y-fronts.

The house was colder, dirtier, quieter . . . and so were the brothers. The Rayburn went out, and they discovered the coalhouse was virtually empty. A desperate search through the kitchen cupboards and the deep freeze was not encouraging, and they soon realised their culinary talents were limited to a series of incinerations.

'Always liked m' bacon crispy anyway,' said Samuel bravely, but he feared with their brand of home cooking they might easily die of salmonella or simple malnutrition.

After a week Thomas dragged the vacuum out from under the stairs, but the bag was full long before he could pick up the clarts from the back porch, and the technology required to resolve that problem defeated him. The widow Jenkins from down the road was hired to 'muck out' two days a week. She prepared a midday meal for them, did some shopping and washing, and charged two pounds an hour. It seemed well worth the money.

Outside, the farm was winding down for winter. Through the short wet and windy days of December the work was done without much enthusiasm. Feed the livestock, put bedding straw into the hemmel, mend a hole in the fence, saw logs, bury a dead sheep. . . . This lack of activity and direction – treading water, and cold water at that – made their situation even worse. No urgency, as there would be at lambing time or harvest, not enough to keep the lads busy, and their minds off their predicament . . . and no income!

It was a month after Ruby's elopement when Mr Thompson phoned again. 'A little chat would be useful,' he said. 'Discuss cash flow, prepare a budget projection, that sort of thing. . . .'

'Cash flow?' asked Samuel, his voice too high. 'What cash flow?'

'Exactly,' said Thompson, sounding like a power-crazed tax inspector. 'That's the problem, – it would appear yours is only flowing one way!'

'I'll ring you back,' spluttered Samuel, 'after I've had a word with Thomas. Ruby used to do the books y' see, and. . . .'

'Tuesday, two o'clock,' said Thompson, and put the phone down.

Samuel and Thomas left for their appointment looking like two 'hit men' from an old Mafia movie. They had unearthed the suits from a cupboard in the spare room and now, brushed down, sprayed with after-shave to combat the mouldy smell and pressed under the mattress

39

overnight, they were ready to astonish the financial world. In fact the lads were ready two hours early, and being 'tarted up', decided to go into town. After all nobody could possibly do anything remotely useful dressed like this. There was a good fire in the lounge bar of the Plough, and they sat there silently drinking until five to two. By then Samuel was as carefree as a well-fed suckler calf, and Thomas was almost asleep.

The bank manager didn't mess about. 'You're over-drawn way in excess of our agreement,' he said sternly. 'What are your current assets?' The brothers looked blankly at each other as Thompson waited, pen poised over a large sheet of paper with a number of boxes marked upon it. 'Same acreage, is it?' he asked. He appeared to be looking down on them from a great height, and his massive mahogany desk seemed as big as a tennis court. Samuel noted the man's immaculate grey suit and Rotarian tie . . . and felt decidedly inferior. He could only grin and nod eagerly. 'Yes, same acreage,' he said, and watched as the banker filled in the appropriate box.

'And what about livestock?' Thompson asked, looking up expectantly. 'I assume you have some livestock.'

'I expect most of them are,' smiled Samuel. A televison screen was flashing the Footsie figures behind the man's chair, and it was difficult to concentrate.

'Most of them are . . . what?' Mr Thompson looked puzzled.

'Live,' said Samuel, and almost giggled. Thomas now had his eyes closed. He was hiccuping, and dreaming of very clever old yowes skipping over the footbath without actually getting their feet wet.

Thompson, quickly realising the interview was slipping out of control, banged dramatically on his desk, and said in a very loud voice, 'SHEEP? How may sheep have you?'

'Two hundred and eleven yowes, four tups, twenty-five hoggs,' said Thomas without opening his eyes.

'What are they worth?'

40

'Fifty quid a head?' suggested Samuel, not very convincingly. 'But if we get a good lambin' they could be worth a lot more. . . .'

A secretary knocked and walked straight in, all legs and eyelashes, but with only one cup of tea, which she placed gently on the tennis court. She smiled at the customers, and glided out again.

'That's hardly good enough,' said Thompson firmly. 'I doubt if you'll be doing another lambing!'

This threat seemed to have a sobering effect on the brothers, and slowly the banker managed to squeeze numbers and valuations from them. The ewes were old; some would not survive through to March, let alone produce twins, – but they didn't tell him that of course. The cattle were sound enough, but very young, probably bought too dear; it would be a while before they could sell at a profit. The machinery was little more than a rusty joke, the values highly optimistic. Trade-in prices might generate some good figures, but that was not a realistic scenario. Hay, straw and a few tonnes of barley added up to almost six thousand pounds on paper, – but if that were realised, then the livestock would have to go as well.

And what about this overdraft? It was climbing at nearly two hundred pounds a week . . . and that was if the lads didn't eat anything or pay the rent. Next year's harvest was a long way off. Thompson concluded there was not much hope. He had already put a receiver into three farming customers, and in one case it had been much too late. Head Office had slapped his wrist quite severely. He was not inclined to make the same mistake again.

Enquiries already made indicated the November rent was overdue (Sir Nigel Nicholas was also a customer at the bank); there was an HP commitment on the pickup, – there would probably be others. The mart was still owed money for cattle bought earlier in the year. Agrigrain had refused to deliver fertiliser, except with payment in advance. Peter Foggin, who was himself in a bit of trouble, had been forced to close all credit accounts on instructions from Thompson. One of these accounts bore the name Forsythe.

It was all a matter of elementary arithmetic now: assets barely covered liabilities, interest was building on interest. Before long, debt would overwhelm them. The only option was to sell up and hope the bank came out more or less even, – and the brothers were left with a few quid, a council house and their pensions.

'Sell up?' exclaimed Samuel. 'Y' mean pack it all in?'

'There's really no alternative,' said Thompson. 'We simply can't lend you any more money. . . .'

'What about Mexico?' asked Thomas. He sounded angry.

'Mexico?' Thompson peered over his enormous desk at the quiet one, and thought he looked very grey.

'Aye, Mexico,' Thomas insisted. 'Y' loaned them millions, – and when they said they couldn't pay y' back, y' told them not t' worry; forget about it, y' said!'

'We rescheduled,' said Thompson. 'This is a completely different situation. . . .'

Thomas had a headache, 'Y' mean Mexico was worth supportin' and we're not. Is that it?'

'I suppose so,' said Thompson. 'But of course these decisions aren't made by me.' (He was using his superior, condescending – 'you may not fully understand' voice.) 'It's Head Office policy . . . farming going through difficult times,' he droned. 'Future uncertain. . . . Mistakes in the past, on both sides,' he conceded generously, shrugging his shoulders, palms up and open. Fortunately he was not smiling, – or Thomas might well have hit him.

They were driving up Rabbit Lonnen to the farm before either of them spoke. 'Is he right?' asked Samuel. 'Are we buggered?'

Thomas, looking straight ahead, said wearily, 'I dunno . . . I think so. . . .'

Samuel pulled up at the back door, switched off the engine, left hand on the wheel, right elbow on the open window. 'Look,' he said, 'if we can get through to the May sales, those cattle could be worth. . . .'

'Don't even think about it,' said Thomas quickly. 'By then we'll owe the bank even more. We're like bloody hamsters in one of those wheels, – the harder y' run the faster y' *have* t' run!'

'But those cattle could be worth three-fifty apiece,' protested Samuel. 'M'be four hundred. Comes to a lot o' money, y' know.'

'Forget it,' said Thomas.

'And a decent lambin' could give us about three hundred lambs at, say, well put them at thirty quid. Fair enough?'

Thomas got out of the pick-up and stood peeing into the nettles. 'Y' can be as fair as y' like,' he said without looking up, 'but Thompson isn't goin' t' sub us any more, – he said so. And what's more, even if we *do* sell the cattle, we'll not be able to replace them again 'cos the bank, the mart, and everybody else will be sittin' on the doorstep waitin' for their money. Some o' *them* are hard up as well, y' know!'

'What about the harvest?' Samuel asked desperately.

'What about it?' said Thomas. 'It belongs t' somebody else! And anyway it's worth nowt 'til it's in the shed. We've always said that, – every time somebody at the Ministry forecasts a record yield in early February.'

In the kitchen they made a pot of tea. 'Alright,' Samuel was resigned. 'So we'd better have a chat with the landlord, I suppose.'

'He'll be pleased to see us,' said Thomas. 'He'll think we're gonna pay the rent at last. . . .'

For several days the brothers hardly spoke to each other. The postman called, and left without his usual cuppa. The back door was open, but Samuel and Thomas were nowhere to be seen. It wasn't the same house without Ruby. Mrs Jenkins came with some groceries, vacuumed the kitchen, cooked a couple of pies and left her bill tucked under a milk bottle. A naive 'Animal

Health Representative' drove in and found Thomas lifting a dead yow onto the link box. 'Lovely morning,' he chirped eagerly. 'Piss off!' said Thomas.

In their private, silent heads the brothers leafed through their memories as they went about the jobs still to be done . . . routine, no enthusiasm now. How had it come to this? What would Malcolm have said? He who had built it all from scratch. What would the neighbours think? Would it be 'serves 'em right', or, 'there but for the grace of God . . . ?' Alfie Sanderson over at Woodlea, he who never made a mistake, he who always topped the mart, he who got four tonnes t' the acre, he who had a computer and a fax machine (whatever that was), – he would say they were too old-fashioned, not on the ball, out of their depth. Maybe he was right. But then again, maybe he was in even bigger trouble, who could tell? Maybe his farm still had a ruddy great mortgage. Or maybe it had been paid off at 5% long ago. Somebody else might be on a rent of forty quid an acre and nowhere to go . . . or maybe his granny left him a fortune and he had no worries. Different farmers, different circumstances, – luck of the draw. How could you tell?

Willie Turnbull at the Glebe would probably say nothing. If pressed, he would still say nothing. If strung up by his thumbs with a fire lit under his bum, he *might* be persuaded to offer something like, 'Well, farmin' was never easy . . . remember the '47 storm. . . .'

In the White Hart, Frank and Arthur Blenkinsop would sit in their corner seats and ponder the Forsthye saga, along with the Common Agricultural Policy, variable sheep premiums, Christmas shopping and the sex life of Gloria Swanson at West Cottage. They might reckon the brothers had simply been unlucky, or hadn't done their sums properly.

Samuel thought about arithmetic as he mixed barley meal and minerals for the cattle. He remembered buying big, rough bullocks for a hundred pounds apiece, and selling them fat within a few months at twice that. He also remembered cattle bought 'on tick' that left nowt. He recalled the tax bill that took your breath away, arriving long after the profits were spent. Those two bad harvests on the trot. Money borrowed. They had never fully appreciated how much borrowed money cost. Money was something you just had to have. The farming wheel had to go round . . . even if you borrowed the axle grease.

Thomas considered the arithmetic as he checked the ewes in Goose Hill, counting them in pairs – 24, 26, 28 – then looking more closely for any signs of trouble, a sheep hanging back, looking dull. Trouble, anxiety, stress, – that was the 'in' word of the decade: stress, a city word, where smart, quick-witted young men were rich and middle-aged at thirty, so they said.

For Thomas, wealth could never be a reward for pressure. It was not an equation that even appeared on *his* balance sheet. All he had ever wanted was his own quietness, his supper on the table, Ruby in the kitchen. A good mule yow with full twins, a fit bullock licking himself, stretching as he got up, a dog that jumped up smiling to greet him every morning.

Additions, subtractions, percentages and profits, – Malcolm used to do all that stuff . . . but he knew (really) that they'd been heading for trouble long before Ruby left. For a start, he had never been happy about credit from the mart . . . too many people knew your business. And Peter Foggin at the garage should have been paid. What was it Malcolm had always told them? 'Even when money's tight, pay the local bills, – not too quickly y' understand, but pay them first. ICI can wait a little longer.' And all those cattle at hundreds of pounds each, it took a fortune to buy them every year . . . the harvest money was gone with no more than a nod to the auctioneer. Quite suddenly it seemed they were understocked, buying poorer quality, trying to spend less, aware of an overdraft that just wouldn't go away. More borrowed, worth less. Like that hamster in the wheel, going nowhere.

'Come bye,' he shouted to Moss. 'That'll do,' and the

dog came happily back to his master. 'Good dog,' said Thomas quietly.

He remembered another piece of Father's wisdom from years ago.

'If you're in a hole and the water's rising,' Malcolm used to say, – 'the first thing y' have to do, is stop digging!'

Yes, it was time to stop digging and climb out of the hole. The question was, how?

By the time Thomas got back to the farmhouse there was just the faintest flicker of a smile on his weather-beaten face. An interesting thought was beginning to take shape.

6

Charles Percival Newcastle-Browne, agent to Sir Nigel Nicholas Bart., sat freezing half to death atop his fat, stagnant chestnut. His whisky flask was already empty.

'The Forsythe boys came to see me yesterday,' he announced. 'I think they may be retiring.'

'What!' said Nickers sharply. He nearly always said 'What!' before and after he said anything else, so Newcastle-Browne did not repeat the news item; he just waited. They were standing by a cluster of bare ash trees, trying to get a little shelter from the east wind, as Ferret the huntsman drew Middleton Wood.

'Selling up eh?' said Nickers. 'Buggered are they, what!'

'Wouldn't be surprised, everybody's buggered these days. . . .' His horse was tap dancing sideways, eager to be doing something awkward.

'Didn't pay the rent last month,' muttered the shivering agent, 'but they're always a bit late. Perhaps we should have a chat about it.'

The hounds were beginning to yelp, their noses onto some foxy smell. Nickers stood up in the stirrups to get a

47

better view. 'Look here Browne, can't talk now,' he said. 'They've gorn right-handed over there. Orf we jolly well go. . . .' And orf he jolly well went. A baronet in pursuit of a brush.

The Nicholas family fortune had come out of the ground, in the great North-East coal-rush of the mid Nineteenth Century. Great-grandfather George had been an engineer; 'Sinker George' they called him, on account of all the shafts he had drilled twixt Tees and Tweed. Before they buried him in the ground he knew so well, he had become the owner of several mines, a country estate, a county wife and a knighthood. His children were educated at private schools to prepare them for privilege . . . and another rural dynasty had been born out of the industrial revolution.

For three generations the Nicholases lived the life of the gentry at Milford Hall. Wars, depressions and a general strike did not seem to seriously inconvenience them. The coal company production rose and fell with demand. Tenants doffed their caps and paid the rents. Local girls found work at 'the big hoose' as kitchen maids, cooks and nannies. The country lads became footmen, butlers, gardeners, grooms. At one time, between Lloyd George and Chamberlain, Milford employed over a hundred local folk on the estate.

There were tea and crumpets on the lawn, horses dancing in the drive. Hunt Balls, with Bucks Fizz at breakfast time, quieter beer-drinking gatherings for the peasants. A Foreign Secretary sleeping in the west wing, a general with a brandy in one hand and a chambermaid in the other. It seemed the sun would never set. But of course it does . . . regularly.

Nigel's father, the old brigadier, having survived everything the Third Reich could throw at him, fell seventeen hands from a demented gelding on Boxing Day 1964, and broke his neck. He was buried in the family plot in Hindburn churchyard.

A fortnight later Uncle Rupert joined him there, having

48

driven his Aston Martin into a stationary oak tree on the way home from a Conservative Dinner at the County Hotel. And that wasn't the end of it. Before the year was out the brigadier's widow carelessly fell off a cruise ship into the Caribbean, – and cousin Bernard, who had always caused the family some embarrassment, was found dead in a bed in Balham. The owner of the bed was understandably upset, but managed to pull herself together and give an interesting account of Bernard's last evening to the *News of the World*.

As the churchyard filled up with unlucky Nicholases the family coffers were drained by death duties. The Government compensation paid on the Nationalisation of coal helped stave off total disaster, – but high taxation and the mounting costs of keeping the family seat warm were making life more and more difficult. Nigel was forced to open the Hall and gardens to the general public on three afternoons a week. On these 'awful days' Lady Daphne would usher fat Americans and shoals of W.I. ladies briskly round the house, and when it didn't rain, thankfully outside, where Ned Green, the gardener, was happy to show off his fuschias.

Two farms on the estate became vacant, and New-castle-Browne advised 'taking them in hand'. Sir Nigel could hire a keen, young, enterprising farm manager who would astound the neighbourhood and generate enor-mous profits directly into the estate funds. Profits which the miserable tenants had been salting away for ages.

Five years later, after some very expensive machinery purchases and abortion in the ewe flock, the enterprising farm manager ran off with an eighteen-year-old groom, several thousand pounds in cash from sales of potatoes, and Sir Nigel's Jag. The farms were re-let. Nickers even considered dispensing with the hounds. To keep the '7th Cavalry' was costing him £50,000 a year. However, when he called an emergency meeting of the Hunt Committee, they were so distraught at the prospect of no more days perched perishing upon their thoroughbreds, they

49

quickly quadrupled their own subscriptions.

And then, as if mother nature was saying, 'you ain't seen nuthin' yet,' one Friday night in October a howling gale blew so fiercely that several chimneys were sent crashing through the roof of the Hall. The damage was considerable, and early estimates to repair it quite surprising. 'What!' exclaimed Sir Nigel when the figure was quoted, 'Old Sinker built the damn place for less than that!'

He knew more money had to be found from somewhere. The Library and several bedrooms were now open to the elements (if not to the public). Five ancient elms had also been uprooted. Three of them had fallen on the stable block, another onto his replacement Jaguar, and the fifth brought down electricity and telephone lines. His London broker rang as soon as he could get through, to inform the unhappy aristocrat that the markets had gone into 'free-fall', and sadly the Nicholas equity portfolio was now worth 40 per cent less than a week ago. 'We've all suffered quite appalling losses,' said the man from the city. 'The Footsie just crumbled overnight, and of course there was nothing we could do about it. The gales, they blew away all the phone lines, the fax, lights out, screens dead . . . really you have no idea, – the wind here in London was extraordinary!'

'It often is,' said Nickers cynically.

He wandered through to the sitting room where Daphne was having coffee. 'Spot of bother, old girl,' he said, looking out of the tall windows, legs apart, hands clasped behind his back, – the at-ease position. 'Seems we might be rather strapped for cash.' It sounded very British and stiff-upper-lipped. The understated catastrophe. Like an officer of the Empire, out of ammunition, his wounded batman and him the only ones left alive . . . the two of them surrounded by ten thousand screaming Zulus. He lights his last cigarette. 'Doesn't look too good, Williamson,' he says, as a hail of spears thud into the wall behind his head. . . . 'Not very promising at all. . . .'

Daphne sipped demurely from her mug of decaf. 'Are we utterly destitute, darling?' she asked.

'What, – well rather. We could sell the Canaletto, I suppose,' said Nickers without turning round. The wind had gone down, it was raining now. 'Might get the chappie from Sotheby's to have a look at it. . . . Keep the Turner I think, don't you, what. . . . Rather fond of the Turner. . . .'

'We won't need to take the boys out of school will we, darling?' asked Daphne. 'I mean it's not *that* awful is it?'

'What, oh good lord no. . . . Of course not. . . . No, no. . . .' He was absent-mindedly staring at the parkland stretching away in front of the Hall, beyond the terraced garden down towards the river. Even on a bleak, wet winter's day it looked gentle, pastoral, permanent. They certainly knew where to build their stately homes, those rich men of yesterday. Sinker had found a wee bit of heaven, and bought it for 'washers'.

'Hindhope!' he exclaimed, suddenly swinging round away from the window.

'Coffee?' Daphne enquired.

'We'll sell Hindhope, that'll do it. Should be worth a few bob, what! Forsythe boys want out, apparently.' Nickers was quite animated. 'Yes, coffee would be lovely, darling.'

What a stroke of luck, he thought, the Forsythes going bust at just the right time . . . marvellous . . . the farm worth so much more with vacant possession, of course. Hindhope was out on a limb anyway, right at the top end of the estate. I mean, if it had been in the middle then one would have had reservations, naturally. Might discover one had sold to a non-hunting man, and damned inconvenient that would be, what! Having to gallop round a bloody sanctuary for foxes. Wonder what it's worth? He set off in search of Newcastle-Browne, playing with figures in his head. He envisaged quite a lot of noughts on the end.

He found his agent over at Woodlea talking to Alfie Sanderson about a proposed rent increase. Sanderson, renowned for claiming vast gross margins, was now trying to persuade his landlord that times were very hard, profits non-existent, and ruination imminent.

Newcastle-Browne was dressed in his agent's uniform: tweed jacket (leather patches at the elbows), calvary twill trousers at half-mast, brown boots and a tartan waistcoat only just meeting across his belly. He leaned on a long hazel stick, the one with a leaping salmon on the horned crook, given to him at the Earth-stoppers' supper two years ago when he proposed a toast to the tenants. He carried it everywhere, like a rural bishop with his mitre.

He enjoyed dealing with the tenants. He knew they didn't like him very much (he seldom brought good news), yet they addressed him respectfully enough, invited him into their kitchens, offered him a whisky. He always called them by their Christian names, rather like a tolerant headmaster dealing with troublesome boys. 'Now then John, Willie, Thomas (or whoever), how are things with you? Behaving yourself, I trust.'

However, this fellow Sanderson was being quite unreasonable; he was actually arguing. These rent reviews were not meant to be negotiable, certainly never used to be. Of course one expected a certain amount of token resistance, – in the good old days even a little pitiful sobbing, perhaps . . . but this chap was talking of arbitration!

'My dear fellow,' smiled Newcastle-Browne, 'arbitration won't do either of us any good. Have you any idea what's involved? Good grief, man, it's just another opportunity for lawyers and valuers to make more money. . . .' He smiled again, as convincingly as he could, leaning as far forward on his shepherd's crook as he dared. 'You'll lose, of course,' he said . . . as if he had inside information.

They hadn't heard the car arrive. Nickers strode into the Woodlea kitchen without knocking, as if he owned the place (which, of course, he did). 'Look here, Browne,' he roared. 'Never mind standing about chatting with Sanderson, – we've better things to do, what! Follow me!'

Newcastle-Browne, who until then had felt extremely important and totally in control of the civilised world, appeared to shrink several inches. His hands slipped off the leaping salmon, and he almost fell flat on his face. By the time he'd recovered some kind of composure Nickers was already on his way out to the car. He hadn't even noticed a smiling Mrs Sanderson offering coffee and digestives on a silver tray.

At Hindhope the cattle in the hemmel skipped about as Samuel bedded them with clean barley straw. The yard had been swept, old tools and instruments, the flotsam that generally lies about, had been tidied away. Inside the house Mrs Jenkins had spent an extra hour cleaning the kitchen and the back porch, and was about to cycle home. The Rayburn, at full steam ahead, pushed cosiness out into the passage . . . the kettle was simmering. Moss barked a warning as the Jag glided in, and then, as if to apologise, went forward, wagging his tail, as Nickers and Newcastle-Browne got out and looked around. 'Thought you said they were broke,' said Nickers, a hint of annoyance in his voice. 'Looks all right t' me, what!'

Thomas came briskly from the sheep pens where he'd been dressing tups' feet, and ushered his distinguished guests into the spotless kitchen. He settled them round the scrubbed table and produced a bottle of Ballantynes with four glasses. 'My brother 'll be with us in a minute,' he said. 'It'll be better if everybody's here. . . .'

Samuel left his wellies at the door and came in as Thomas was pouring the whisky. 'Your astonishing good health, gentlemen,' said Nickers, raising his glass and beaming at everyone. 'Cheers, what!'

They discussed the weather, how short the days were, the price of hoggs and store cattle, the number of foxes about; Thomas with his rear-end to the fire, Samuel sitting opposite the two visitors. 'Well then, boys,' said Newcastle-Browne taking advantage of a pause in the chat. 'What seems to be the problem? Rent, is it? Really should be paid by the end of the week at the latest. . . .' He was feeling important again, puffed up.

His boss deflated him immediately. 'Nonsense,' roared Nickers, 'we're not here as bloody rent collectors!' He took command of the whisky bottle, poured a little into the brothers' glasses, ignored his agent, and gave himself another generous measure. 'Believe me, I understand the problems only too well,' he said. It was his enlightened Landlord's voice, compassionate, full of understanding. The same tone Lady Daphne used when she looked down from the magistrate's bench on some unfortunate unmarried mother of five kids who'd been caught shoplifting in Presto's.

'Oh my word yes, fully appreciate the difficulties of farming today,' he went on. 'Tried it m'self for a while y' know, – damn glad to be out of it, what! And of course at your age' (the practised on–off smile) '. . . all those years eh, changing . . . ever-changing. . . .'

Samuel thought he sounded like the vicar, but perhaps it was only the mellowing effect of the drink.

'To be honest,' said Nickers, 'I'd pack everything in as well if I could. Absolutely, tremendous pressure these days, what . . .' He emptied his glass, then refilled it with what was left in the bottle. 'In fact, I don't mind telling you, when you boys retire at the May term we intend to sell Hindhope . . . ease the load a little . . . far too much on m' plate. . . .'

Thomas eased himself away from the fire. 'Aye well,' he said, looking down at his feet, surprised to see a big toe peeping out of his wellie sock, 'it's . . . er . . . not quite as simple as that, Sir.'

'No, it's not that simple,' Samuel chipped in, without any idea what his brother was on about.

Newcastle-Browne, eager to be involved somehow, desperate to re-establish his credentials as the 'Grand Vizier' of the estate, pushed back his chair, looked pointedly at his watch and asked, 'have we a problem here, gentlemen?'

'Rubbish!' exploded Nickers. 'What, – there's no damned

problem at all. . . . The Forsythes bugger orf in the spring, have a splendid sale, buy a nice little house somewhere, live happily ever after, and we sell the farm to some poor sod with more money than marbles, what!'

He laughed enthusiastically, looking round for signs of approval, and banged his empty glass on the table. Samuel went to the press and came back with a dusty bottle of Bells, with no more than a couple of measures in it. It was the last of the lamb reviver, left over from March.

'More money than sense, eh?' said Thomas. He had managed to surreptitiously pull down the end of his holey sock and conceal the naked toe. It made him feel less vulnerable. 'If we pack in,' he went on slowly, 'you can sell at vacant possession and make a fortune . . . but where does that leave us?'

'Retired,' said Newcastle-Browne. 'Feet up, no worries, sit in the sun. . . .'

'Suppose we don't,' said Thomas.

'What?' Nickers could feel apprehension creeping up his jodhpurs.

'Suppose we don't bugger off?' Thomas had sat down opposite the landlord. 'Suppose we stay put; where does that leave you?'

'Ah well now,' said Nickers thoughtfully. 'Perhaps we do have a teeny weeny problem after all, what.'

'It's the rent, isn't it?' Newcastle-Browne interrupted quickly.

'Oh shut up about the damned rent,' shouted Nickers. 'The rent's of no consequence whatever. . . .'

Samuel was surprised to hear this, but said nothing. He was watching his brother.

Thomas poured the last of the Bells into Nickers' glass and said simply, 'you see . . . we may be *forced* to struggle on somehow. The fact is, I'm not sure we can *afford* to retire!'

Nickers looked at him warily for a moment or two . . . and got the message. 'Oh yes you can,' he said quietly, 'I guarantee it!'

7

The news was officially confirmed in the White Hart at 8 o'clock that night. By closing time it had spread to four other pubs, and by noon the next day the whole parish was privy to the plot.

No one was quite sure where the original leak came from; certainly not from Samuel or Thomas, they hadn't even ventured out. Somewhat apprehensive about their future, they had sat all night at the kitchen table wondering, worrying and drinking tea. Nickers and Newcastle-Browne were closeted with accountants until well after dark, before retiring to the County Hotel for Dover sole and three bottles of Chablis.

Jack, the publican, said the first he'd heard of the rumour was from the council bin men who had come in for a pint and a toasted sandwich at lunch time. They had definitely been discussing a farm for sale *somewhere*, but hadn't actually mentioned a name. Jack had thought no more about it, until Geordie Dodds and Willie Turnbull came in about seven. Geordie said, 'I hear his Lordship's hard up, lost a fortune in the city they tell me . . . might have t' sell some land.'

Willie said, 'aye, – maybe . . .' which didn't really add much to the debate.

A quarter of an hour later Alfie Sanderson came through the doorway and asked if anybody had heard the story that Nickers was giving up the Hunt . . . couldn't afford to keep it going.

At 7.30 Percy Elliot, the mart secretary, entered with two noisy cattle dealers from Easingwold. When asked if *he* had heard anything, Percy withdrew into his conspiratorial shell, denied all knowledge of any farms for sale *anywhere*; but denied it in such a way that left his audience suspecting he knew far more than he cared to admit. Given the nature of his job, he often did. The two dealers said they'd buy it *now*, for *cash*: Where was it? How big? How much?

It wasn't until Sep had ordered a pint of Special, and downed half of it, that the rumour began to acquire any credibility. He cleared his throat, a dark rumbling device he always used to prepare for some pronouncement. Sometimes it proved less than profound, sometimes it was so blindingly obvious, it surprised everyone. This time it stopped the chatter in its tracks.

'It's Hindhope,' he said.

'How the hell do you know?' somebody asked.

'It's right enough,' said Sep quietly. 'The Forsythe lads are retiring, and I hear Nickers is gonna sell the place. . . .'

'Get away. . . .' said Willie Turnbull.

Somebody over by the dart board said he didn't believe it. 'There's been nowt in the papers,' he argued.

'Please yerself,' muttered Sep. 'It just happened t'day. . . .'

'Where did y' hear this, then?' asked Geordie. 'D' y' reckon it'll be true?'

Sep cleared his larynx again, even though it wasn't really necessary. 'A pretty reliable source,' he growled confidently, and ordered another pint.

'Who?' demanded Jack from behind the bar; 'where?'

Sep waited until he felt they were all looking his way and, without raising his eyes or his voice, said, 'A certain postman told me. . . .'

That did it! That was the confirmation the known civilised world needed. If Harry Bell said the Forsythes were coming out and Nickers was selling Hindhope, it would be 'gospel'. That man knew everything about everybody . . . sometimes before they did. Of course there was never any suggestion he steamed open the Royal Mail with a portable kettle carried in his van . . . nothing remotely criminal or devious like that. As he himself said, you didn't have to be the local CIA agent to know the difference between an ordinary invoice and a final demand.

It didn't take a bloodhound to sniff out a clandestine love letter or a grain cheque. If anyone had asked him, there's little doubt he would have been able to determine the contents (near as damn it) of almost any package he handled. This talent was not acquired overnight, – something near thirty years on the early morning run, a cup o' tea here, a bacon sandwich there . . . half-heard kitchen arguments interrupted from the back door . . . 'Anything for the Postie?'

Yes, if Harry said the story was true, you could print it, and by the following night it was cool news already. Everybody knew.

There were to be two sales. Samuel and Thomas, having made the big decision, or more realistically, had it forced upon them, – concluded the quicker they got out the better. Quit while they at least had *some* control in the matter. Nickers, for reasons not entirely altruistic, had offered to waive any rent arrears, provided the brothers were gone by May. Futhermore, when pressed by Thomas, he had promised one of his cottages at the west end of the village at a peppercorn rent for the rest of their lives. Samuel would get his full pension next June, Thomas two years later. Then, if they had a reasonably good sale, plus the 'outgoings', they could pay off the bank, the mart and the rest, and perhaps have something left over to put in a Building Society.

They were never going to be 'well off', but neither of them wanted a fancy lifestyle anyway. Ascot, Wimbledon, a yacht in the Med meant nothing to them. If the pick-up passed its MOT, that would do nicely. If they could have a jar at the White Hart now and then, and swop rural gossip, – fine. Mrs Jenkins would 'muck out' on Mondays and Fridays. Then, – no more lambings on cold, wet nights, no more harvesting after everyone else had finished, no more worries about droughts and floods, a sick bullock, a poor price at the mart, lambs rejected, bills unpaid . . . the ominous ring of the telephone. Ah yes, the prospect of retirement was beginning to look quite attractive.

The mart company advertised the Hindhope sale for April 24th. Buyers would have some spare cash by then. Livestock prices would be picking up with the first signs of spring grass.

So, there *was* one more lambing after all, and the old yowes (having done it all several times before) played a blinder. The weather was kind, Thomas fed the sheep well, and ended up with a genuine 173% . . . if you counted the three pets . . . and forgot about the four yowes that died.

In the days running up to the sale he had sheep in the pens every day, – combing, clipping and titivating the ewes like a rustic Vidal Sassoon, sorting them into handy saleable lots. There were also half a dozen geld yowes (as fat as seals), three creatures with damaged 'under-carriages', four arthritic tups, and the twenty-five wether hoggs who never made the grade last year (most of their wool had fallen off, but they were quite fit). Everything had to go.

The cattle looked fairly good, Thomas thought, – especially when drawn into matching groups and trimmed with the shears. In a small, tight ring they would look even better.

Samuel busied himself among the deadstock. With spanners and oily rags he tidied up their antiquated collection of gear till even the rust looked better. The Massey got fresh oil filters and a battery change. Everything had to go. The old cultivator had seven new tines fitted, a broken shaft was welded on the Cambridge rollers, a spare knife found for the reaper. Everything was dragged out into the Croft, numbered and catalogued by the auctioneer's clerk. Spades, forks, a ladder, wire netting, a wheelbarrow, posts, rails, hammers, a box of nuts and bolts, a bird scarer with gas cylinder, sheep and cattle medicines (with expiry dates reaching back to the middle ages), earthenware troughs, hay hecks, some horse harness . . . and all the other farm debris washed up on the tides, were now lined up for display. Everything had to go. Everything that is except Moss . . . Moss would retire to the cottage with the lads.

Meanwhile two busy young men from Spratt and Mackeral's Harrogate office had begun their campaign to sell the farm. They came armed with cameras, clip-boards and CLA ties. They marched purposefully over the fields, through the buildings, peering, prodding and measuring. They drew up a comprehensive list of adjectives: *Productive, attractive, traditional, imposing, impressive, picturesque,* – even *spectacular, unrivalled, unspoilt* and *unique.* They sat with Samuel and Thomas (*quaint*) drank their coffee (*disgusting*) in the farmhouse kitchen (*medieval*) and eventually boarded their Range Rover (*essential*) and headed back to HQ to compose the glossy schedule.

On the cover was a panoramic view of the farm taken from Rimside field. The photograph was in colour, snapped on a crisp March day, the shadows sharp, the stones cool: an impression of rolling countryside laid out around the farm like a ruffled quilt on a comfy bed.

Open it up, and the plot was clear in bold type from page one.

HINDHOPE FARM, MID NORTHUMBERLAND, a renowned

arable and livestock holding located in a productive area of the county. Substantial farmhouse, range of traditional buildings, 245 acres. For sale by private treaty as a whole or in lots as required. The property is offered freehold with vacant possession upon completion.

Over the page, and carefully worded paragraphs talk of A rare opportunity to acquire a quality farm in attractive and fertile area with easy access. Not too easy you understand, better to preserve an attractive rural image. But the map on the back cover showed the A1 only twelve miles away, and there in the bottom right-hand corner, the edge of the city with international airport, providing regular services to London Heathrow/Gatwick, and BR's intercity link to Kings Cross in under three hours.

And this is hunting territory, – there is a river, and as the blurb says, It is considered there are further opportunities to expand upon the strong sporting potential of the farm.

On page three, the farmhouse, with benefit of superb south-facing location, is pictured, together with detailed descriptions and measurements of her innards. Kitchen: 20' × 17' 5". Rayburn cooker serving hot water. Range of wall and base storage cupboards, double drainer sink. Pantry 10' × 6' fully shelved. Living room 16' × 14' open fireplace. Bedroom 14' × 12' Victorian fireplace. Bathroom 9' × 7' with cast-iron bath, basin, WC and linen cupboard.

Not a mention of any double fitted wardrobe with a centre vanity unit. No power-shower cubicle, no bidet, no central heating radiators. It was almost a surprise to read there were mains electricity and a telephone. A challenge perhaps for some country-loving couple with the money to match their imaginations.

The garden got the briefest of mentions. Walled and mature, said the script, and hurriedly moved on to the other things. The farm buildings incorporate a traditional range of stone-built byres and loose boxes, part with lofts

*over, with slate and pantile roof, around a courtyard to
the rear of the farmhouse . . . which, subject to the grant-
ing of planning consent, are considered suitable for con-
version to residential units if required.*

On page five, another panoramic picture attempting to
be arty. The lens peers through branches in the fore-
ground towards an open gate, and out across a meadow.
Some cattle appear to be advancing towards the camera
(presumably curious to meet the idiot who left the gate
open).

The land is in exceptionally good heart, we read, *well
fenced, drained and watered throughout. The farm lies
approximately 500 feet above sea level, and much of it is
cropped in a traditional rotation. Access to all fields
is exceptionally good, having benefit of extensive road
frontage.* It goes on . . . *under Ministry of Agriculture
Land Classification Schedule, Hindhope is identified as
being Grade 3, but it is felt that parts of the holding have
the potential of Grade 2.*

There is more, – and it's all good news of course . . .
*Woodlands provide excellent shelter and sporting poten-
tial . . . they contain some mature hardwood capable of
producing immediate sale income. . . .*

Further words on Boundaries, Sporting rights, Mineral
rights, Rights and Easements, Disputes, Arbitration; the
map with O.S. numbers and exact acreages (Wood Close
13.835 acres) divided into arable, temporary and permanent pasture, to the bottom line on the last page:
*Guide price: Offers in excess of £300,000 are invited
no later than May 13th, and prospective purchasers are
advised to register their interest at an early date.* You've
read the book, – now see the real thing!

The brochure went out to everyone on the Spratt and
Mackeral mailing list. Adverts were placed in major
agricultural publications and in the property sections of
most newspapers, national and local. Anyone looking for
a farm could hardly miss it.

The Forsythe boys watched with a mixture of sadness,

resignation and a degree of embarrassment as the notice went up at the farm gate for all to see. They knew their independent world was now open to inspection by the public, albeit 'By appointment only'.

Most farmers tend to live rather secretive business lives. Profit and loss, details of rent and bank borrowings are generally a closed book . . . sometimes even to wife and family. Good news shouted about only tempts providence; a dead yow is best forgotten anyway. Samuel and Thomas had lived that private life. It simply wasn't anybody else's business exactly how many lambs were born or how much barley was harvested (always provided the figures matched the other half-truths and exaggerations bandied about at pub and mart). Their space had been private too. Hindhope was *their* farm (even if they didn't actually own it). *They* decided who came into the yard, walked over a field, sat on a fence and passed the time of day.

Now they would come by the car load, the curious and the keen, the neighbour and the stranger. The successful farmer with an eager son, a rich townie with pastoral delusions and a green, persuasive missus. There was nothing the brothers could do about it now.

It was Thomas who suggested they should move out as soon as possible. He said he didn't feel the house was theirs any more, – people wandering about the fields and buildings every day as if they had taken over already; paddling along the passage, upsetting Mrs Jenkins, using the toilet, opening cupboards, even looking under his bed. The cottage in the village was in far better repair than the farmhouse anyway; all it needed was a good scrub, the electricity switched on, the windows opened.

So Samuel began moving furniture with the pick-up, and a fortnight before the sale they were settled in, carpets on the floor, a fire in the grate, food in the fridge. They were astonished how warm it was, and with the White Hart only a few yards away, it was going to be all right. It was cosy.

One night, just before the big day, with everything prepared up at the farm, Thomas, Samuel and Moss were watching television, when there was a knock at the door. Moss considered barking, but he wasn't all that confident yet of his new, improved circumstances behind the sofa, so he just looked fierce and laid low. It was Thomas who eventually groaned to his feet, and went to see who it was.

Ruby stood there smiling.

Thomas took her bag. 'The door wasn't locked,' he said. 'Y' should've just walked in. . . .'

8

Gerald was extremely weary. On Friday nights he was especially extremely weary. For five days he had fought his way through the traffic to the office, competing with salesmen in Sierras, managers in Mercs, and a procession of Polos and Golfs in which fraught mothers carted their dull early-morning children unwillingly to school. It was almost as bad coming home again, – except the mothers had gone by then. At 6.00 p.m. there was just the mad one-way stampede into suburbia, onto the motorway, out of the city . . . and home.

Gerald didn't want a kiss or conversation; he wanted a gin and tonic, and then another gin and tonic, and maybe a third. Give him an hour or so for the telephone in his head to stop ringing, his intestines to rearrange themselves comfortably, his upwardly mobile metabolism to cool down. Switch off, shoes off, old sweater, the *Independent*, the end of the week . . . sink slowly in the nest.

But tonight Prudence would not allow him to settle. His drink was ready as soon as he walked through the door, – and so was her speech. It was one of her mobile speeches, given at different noise levels, as she flitted from cooker to sink, from kitchen to lounge. He'd heard it all before, or at least something similar. He occasionally missed a word or two if the tap was turned on, – or if he was turned off.

'It's all about quality of life, darling,' he heard her say (yet again). Why, he wondered, did they call each other 'darling' all the time, – even when they were having an argument. He briefly imagined himself strangling her slowly with his National Trust tie, and asking with an evil grin, 'Had a nice day, darling?'

'The Prendergasts were burgled again last week,' she said. 'All the usual stuff, – telly, video, jewellery, some lovely silverware . . . and that oil painting Abigail claimed might be an early Frith, though no one could ever find a signature, of course. . . .' She was laying the small table on which they always had supper when alone. 'And Neville Potter had his car stolen . . . third one this year. . . . Actually saw them drive it away, – three youths with rings in their ears. It's anarchy,' she said. 'They'll probably pinch the radio, and then push the car over a cliff. . . . Can I freshen your drink, darling?'

He pictured himself hurling her off Beachy Head and saying, 'Whoops, – sorry, darling.'

She disappeared into the kitchen again to check the simmering veg, and the voice went down a few decibels. 'What was the traffic like today?' he heard her ask (she didn't expect an answer). 'I met Janet for coffee in town and it was horrendous. Couldn't get parked . . . and the carbon monoxide, my God there is simply no breathable air left on the planet!' She came through to the lounge with a glass of Presto's special-offer Australian dry white and a glossy green brochure. 'Supper in five minutes,' she announced, '. . . and oh, by the way, I picked this up. . . .'

Gerald didn't even look at it. 'Not another one,' he groaned; 'how many have you got now?' It wasn't really a question, more a criticism of the growing collection. He was feeling better though, – the second large gin had gone down nicely. He even began to feel just a little guilty about his attitude.

As if on cue, Prudence said, 'Your attitude's totally unreasonable, darling, – you just dismiss the whole idea, no discussion. . . .' She went back to the kitchen and banged a few pots and pans.

'All right, I'm sorry, darling,' he shouted after her. 'But really only an idiot actually *wants* to farm these days, you know, – agriculture's bloody bad news. I mean it's all very well for proper peasants who've been there for yonks, – they know no better, poor things . . . but it's not a sensible option for *us*, dear . . .' She carried the potatoes through to the lounge, went back for plates. 'It's bad enough *living* out there,' he went on, 'but messing about with mad cows in the middle of a field . . . well it's so damned . . . (he struggled for the word) . . . so damned rural!'

Prudence served the casserole, poured herself another white wine and sat down. She knew his views on the subject well enough. Here they were (he'd say), very comfortable in Acacia Avenue, not rich you understand, not in a serious mega-bucks way, but – pretty well off, holidays in France every year, the house worth quarter of a million quid perhaps, with a very reasonable mortgage. Two cars in the garage, two kids at school. She saw a fleeting image of Julian and Emma running and laughing through a field of buttercups, with big, brown animals grazing there; she sitting on a gate as the children came towards her. They looked so happy and healthy. 'Healthy and happy . . .' she murmured to herself.

They ate silently for a while, or at least without speaking, though the noise of cutlery and crockery, even swallowing, sounded exaggerated.

'That was delicious, darling,' said Gerald pushing his empty dish away. Prudence was still sitting on her gate, the sun setting behind her, not a cloud in the sky. She was idly moving her supper around the plate.

'Look, darling,' he said in his infuriatingly reasonable voice, 'we know absolutely nothing about farming. It's a high-tech business now, or so I'm led to believe; and, well, I'm not sure we can afford it. Decent land costs about two

67

thousand an acre, I think . . . or is it a hectare? Anyway it's a hell of a lot. And that's just for starters; I expect we'd need some sheep and chickens and cows . . . a tractor . . . that sort of thing. Various odds and ends. . . .'

'So you have at least been thinking about it,' Prudence said coolly.

'Oh of course, darling . . . I realise you've had this silly idea for some time now.' (she scowled at him); 'all right then, this picture of the idyllic life filled with fresh air and friesians, a return to nature, rediscovering oneself, and all that stuff . . .'

'We *could* afford it,' she said.

'But you know nothing about the countryside!'

'I can learn. Women like me do Open University degrees in their spare time, Computers, Law, whatever . . . God, feeding a few sheep can't be all that difficult. And anyway I'm sick to death of Save the Children coffee mornings, aerobic classes with wobbly wives, and vacuuming this house . . . and the whole worthless existence I lead. I'm a country person, I want to be a *farmer*!' She banged the table and the dishes bounced. Gerald bounced a little too. 'If people like George the Third and Michael Jopling can do it,' she declared, 'anybody can!'

He didn't know what to say, so he said what he always said when she was on high doh, when she was unlikely to be diverted or appeased by anything compelling . . . like logic; he said, 'of course, darling . . .' and began to do the washing up. Prudence sat with both hands around her wine glass and let him get on with it. This time she was determined to run all the way to the tape.

Later, in bed she said, 'so what *are* we worth?'

'Difficult to say,' Gerald looked at the bedside clock. It was eleven thirty. He closed *A Year in Provence*, somewhere in mid March, at the point where a poor peasant had lost a third of his melon crop, consumed by wild boars. 'The house? Maybe £200,000,' he said. 'Equity in the company perhaps fifty K, a few investments here and there, unit trusts . . . pay off the mortgage. . . .' He was doing sums in his head while still visualising pigs stuffing themselves with melons. 'I dunno . . . three hundred thou,' he said, 'if we realised everything.'

'That would do it,' said Prudence. 'For that sort of money we'd get a beautiful house, lots of lovely old farm buildings, and 240 acres of land!' She was appealing to the businessman who lay next to her, scratching his belly. She said, 'I mean for the future, for the children, for our old age. You can't beat land as an investment, can you? Who was it said they aren't making it any more?'

'Mark Twain, I think,' said Gerald staring at himself in the dressing table mirror, 'or Oscar Wilde, perhaps. It's always one of those two . . . but neither of them were farmers!'

'Well, doesn't it make some sort of sense to you?' she asked. 'Value for money, privacy, independence. . . .'

'Farmers never make any money.'

'Of course they do,' said Prudence sharply, 'they only pretend to be destitute. What about that friend of Neville's, that incredibly uncouth chap we met there last month, went on and on right through dinner about ewes having prolapses . . . disgusting. *He* was a farmer, drives a BMW and plays golf three days a week, – enormous hands; on Clarissa's thigh most of the night, I recall.'

'Good Lord,' said Gerald, 'really?'

'Hindhope,' said Prudence, 'is in a perfectly beautiful part of the country. You could commute to work easily, and I'd look after the farm. What's more,' she said, 'I honestly believe Julian is a country boy at heart, I think he'd make a jolly good farmer some day. Let's face it, he's never going to be an academic, is he? He loves animals, – he could take over when it all gets too much for us, and we could retire and stay there for ever. . . .'

'You seem to have it all worked out, darling.' Gerald had slid down under the duvet; he wished she would switch off the light.

'Emma could have her pony,' she said. 'You could help at weekends, it would be good for you . . . therapeutic. We

69

could send Julian to Cirencester. He could learn to drive a tractor, milk a cow, that sort of thing.'

Gerald's eyes were closing. 'So you want us to sell the house,' he moaned in mid yawn, 'liquidate all our assets, borrow some more cash, and pour it all down this rustic drain. Then we all live happily ever after as impoverished peasants, subsidised by my proper day job. Is that it?'

'Don't be ridiculous, Gerald,' she said impatiently. 'There are all kinds of EEC grants and subsidies; we'll make lots of money . . . Gerald, you're not asleep are you?'

'Yes,' he said wearily, counting the melon-eating pigs as they trotted off into the dark, dark undergrowth. . . .

On Saturday morning Prudence phoned her mother, as she did every weekend, and told her she had picked up another 'Farm' brochure. Granny phoned her other daughter, Mildred, to tell her Gerald and Prudence had finally chosen a farm. Mildred told her husband, Mildred's husband informed his cousin Michael and, when cousin Michael met Gerald on Sunday morning on the first tee, the first thing he said was, 'I hear you've bought a farm, old chap. . . .'

'I most certainly have not,' Gerald protested, and promptly hooked his drive into the car park.

That evening they were out to dinner at the Nicholsons'. Hugh, a solicitor, generally had a fund of sordid legal tales to tell, disorderly goings-on in semi-detached suburbia, juicy extra-marital shenanigans beyond belief. He could be very entertaining. The delightful Delia would produce a splendid meal, and float about in some diaphanous creation looking like a nymphomaniac moth. Should be a good night. Apparently the Prendergasts were going too, – excellent. Larry was a wine merchant, and with a bit of luck he would bring along something interesting to compensate for the Rumanian rosé Hugh got cheap from his underworld connections. Abigail Prendergast would undoubtedly be drunk before the main course, and embarrass somebody. Yes, it could well be a very promising evening.

Neil and Marcia would be there. Neil worked abroad most of the year, – Nigeria, Libya, desperate places like that . . . something in oil, wasn't he? A real tycoon apparently, with a very pretty wife, who, rumour had it, became very lonely while he was away. All sorts of interesting gossip about her. Prudence didn't like her at all. Gerald was definitely looking forward to dinner at the Nicholsons': good conversation, amusing company, his kind of people . . . city people.

They were the last to arrive, and had to park in the street. The door was open; warm light, sounds of clinking glasses and laughter drifting out through the hallway . . . inviting. Delia fluttered forward to greet them. Hugh, the perfect host, kissed Prudence on the cheek, took her coat, said how lovely she looked, and thrust a gin and tonic (iced and lemoned) into Gerald's outstretched hand . . . splendid. Ushered through to the lounge, lots of 'hallo darling, haven't seen you for ages. . . .' Everyone talking at once.

'Listen everybody,' shouted Hugh. 'Guess what, Gerald's bought a bloody farm!'

'Oh, how terribly exciting,' purred the tycoon's wife, gliding closer. 'A rural retreat eh? What fun . . . all those rams and bullocks making whoopee in the grarss. . . .'

'We haven't bought anything yet,' Gerald protested desperately.

'But we are thinking about it, aren't we, darling?' said Prudence.

9

Les Stevenson started business as the local 'odd-job'. He was the lad who rebuilt your garden wall after some drunk had demolished it, or put the slates back on the roof after a gale. He was the fella who assembled the new greenhouse on a flat concrete base, laid the patio, installed pine kitchen units. All for cash of course, – none of this silly VAT nonsense. He was available Saturdays, Sundays and Bank Holidays, and usually ate his bait standing up.

After a couple of years he had a Toyota van and called himself a 'builder'. Before long he was employing a plumber, a plasterer and a chippie. They did the extension over the garage for Mrs Williamson's mother in 1976, began renovating old cottages for the council, – and built their first complete house two years later.

Les went from strength to strength. He bought plots of land here and there, purchased some derelict property, and when the housing market took off in the mid eighties, he was sitting on a little goldmine.

Mind you, – it did his cause no harm to play snooker

every Friday night with Charlie Woodruffe, the Borough Planning Officer. He joined the Rotary, the Lions, drank at the Conservative Club, and could always tell a bad joke very well. 'A bit of a character,' they'd say, 'done well for himself. . . .' Then they might add thoughtfully, as if to allay any doubts there might be regarding his integrity: 'canny bloke, though.'

Les would never claim he was the first to recognise the residential potential of old farm buildings, – but he was certainly pretty quick off the blocks.

Alfie Sanderson had bought a big marginal farm west of the A1. It was about a thousand acres, not too dear, and attracting a lot of subsidies. It had been the country estate of a city industrialist who, having 'played' at farming for a while and made a few mistakes, decided he would be far better off (and much happier) sticking to the manufacture of nuts and bolts . . . or whatever it was his factory churned out at Gateshead.

However, in the early days of this agricultural adventure, he had been advised that what the farm needed was a new up-to-date steading, – and his bright young farm manager (with a B.Sc. Agric., clean fingernails and four years experience with a worm drench company) did little to dissuade him. Consequently up went a massive sheep house, complete with dipper, footbath and pens – concrete all over the place – and another building for cattle, that would have nicely accommodated a Rolling Stones concert. The old steading which, fair enough, had lost most of its appeal for both man and beast anyway, was simply abandoned . . . along with the tumbledown lodge at the bottom of the road, and three nineteenth-century farm cottages with outside netties, stone pigsties and see-through roofs. They were all left to talk among themselves of better days.

Sanderson stocked the land with an assortment of suckler cows put to a cross-Charolais bull, a lot of Blackie yowes who did their level best to produce a mule lamb apiece . . . and sent his eldest son Jimmy to live there and run the show. He gave him a secondhand Land Rover, a Calor gas cooker, one of those ATV trikes, a useful collie dog, and advised him to get married as soon as possible. After five weeks of supermarket soup and sausages, and in spite of a meaningful relationship with the dog (who slept under his bed), Jimmy quickly declared his desperate devotion to a lass from the library. She, being two months pregnant, readily accepted almost before Jimmy's bending knee had touched the grass. She moved up to the farm immediately, kicked the dog out into the yard, and began to redesign the kitchen. That's when Les was called in, and got his eye on the derelict buildings.

He offered the Sandersons £150,000 for the byres, the ruined lodge and the three old cottages that stood in a line halfway up the farm track. Alfie, who had maybe stretched himself a wee bit and was watching bank rates with some concern, shook hands on the deal without much delay. It meant he was left with the farm for 'next t' nowt'.

With the co-operation of Les's snooker pal, Charlie, and the chairman of Rotary (who happened to be chairman of the housing committee as well), a scheme for five homes was duly approved. The Parish Council made some noises about drainage and septic tanks, but work began within weeks. Long before the roof was replaced on the lodge, a retired couple with a married daughter in Newcastle, and a lot of spare cash from the sale of their semi in Wimbledon, had signed on the dotted line and paid the deposit.

A barrister with two BMWs and three children at school in Durham eventually bought the three old cottages, which Les Stevenson Ltd, had transformed into a 4 bed, 2 bath Dallas-type bungalow, with paddock. It cost the lawyer over £200,000. The old barns and byres were virtually razed to the ground, to re-emerge as 'three charming executive dwellings.' They were quickly snapped up by three charming executive couples (each with two salaries and no kids). One pair even paid cash. Les was quite pleased. It had taken less than two years.

When he saw the sale notice for Hindhope he promptly sent for particulars and phoned Charlie Woodruffe. 'I realise it's not Friday night,' he said, 'but do y' fancy a couple o' frames?'

* * *

Someone else who read the sale signs with more than a passing interest was Sep Robson at Clartiehole. The Robsons and Forsythes had been neighbours all their lives, and their fathers before them.

Sep was the archetypal umpteenth-generation peasant, – built entirely with country chromosomes; the image most townies had of how a proper farmer should look; the picture children's story books painted. A solid man, substantial, difficult to move aside, feet and hands that looked as if they might have been grafted on from someone slightly bigger, a face the colour of autumn, – except beyond the cap line. A cap that only came off at the very end of the working day.

He was a farmer who wore his emotions on a rolled-up sleeve. If he was busy, visitors were often invited to bugger off. If things were going well, he might stand and blather for an hour. He was a stockman of considerable talent: a good judge of cattle that might grow into money, a shepherd who saw trouble early, who would get the best out of a lambing, – given any luck at all. If he had a weakness, it was a lack of patience. 'If you're goin' t' panic,' he would argue, 'for God's sake panic quickly!' And so sometimes the hay bales were a touch heavy, – occasionally the barley would have been better left another day. Sometimes he knew he should have said nowt.

Gladys had been with him for thirty years, and she definitely knew when to say nowt. You simply did not ask what the problem was, – at least not until the problem was resolved. When she heard him miles away, screaming obscenities at the dog, she would resist the

75

temptation to criticise his vocabulary until the dog was fed and locked up for the night. If he walked into the kitchen with only the top half of the pet lamb's milk bottle in his hand, – she knew that some inane question such as, 'wouldn't it suck then, dear?' could produce a disappointing reaction. She knew him well, he was not a complicated man. They were on the same team.

The children? Well, they were hardly children now, – 'bloody near middle-aged,' said Sep. But they'd been little bother . . . not when you read about the way some kids behave.

Willie was a farmer from square one, – couldn't wait to get home from school and into his wellies . . . followed his dad around the fields like a second dog. He could lead corn off the combine without spilling a pickle, and reverse a big trailer into the shed without hitting anything, long before he had a licence. Sep *still* couldn't do that, but then he would be the first to admit he was never really 'mechanically minded', like the young 'n's. Now all of a sudden the lad was twenty-six and courtin' strong . . . Sandra, bonny lass, worked in the Halifax.

Daughter Doreen was married already to a canny hill farmer up country. Gladys always referred to her as Mrs Armstrong now. They would meet on Saturday mornings for coffee and do some shopping. 'You're not expectin' yet, Mrs Armstrong, are y' ?' she'd ask, – hoping she was. Gladys fancied being a granny something terrible, and made little attempt to conceal it.

Meanwhile Sep was eyeing Hindhope with more than just the detached curiosity of a bloke next door. He knew the farm well enough of course, he had compared disasters over the dyke with Thomas on many a morning. He'd often been over there to retrieve a few nomadic yowes or a wandering heifer, or to swop machinery. He gave the boys a helping hand to lead the last of their hay under cover when the forecast was ominous (but only if his own was already in the shed). They were good neighbours. He was not delighted to see them going downhill, but it concentrated the mind on his own dilemma. . . .

Clartiehole was just over three hundred acres, and had provided for the family well enough through the peaks and troughs of farming. There was some good land on it, – you could expect a three-tonne winter barley crop now. Most of the lambs went away fat, straight from their mothers. Sep could still make a bob or two buying back-end store cattle, wintering them rough, and selling them off the grass. However, it was all becoming harder. Very soon Willie and his wife would require a living from the place as well. Clartiehole was not big enough for two families. The day of the small, self-contained family farm was coming to an end. Sep had thought that for a while now, – he'd seen some of them disappear already.

So what about Hindhope?

Well it would go well with Clartiehole, no doubt about that. But he couldn't afford to buy it all on the open market. You could bet your wellie socks there would be some silly money about, and the average tenant farmer couldn't hope to compete, unless he'd been bequeathed a fortune from somewhere. Sep had no such expectations . . . and he certainly was not about to borrow himself into a bottomless pit.

There was just a glimmer of light . . . a small window of opportunity (as the politicians say). Sep looked at the sale notice again. 'For sale as a whole, or in lots as required,' it read. Clearly Nickers had been advised by Spratt and Mackeral that there could be more money if the farm was split up. The bricks and mortar sold separately, perhaps . . . an agri-asset stripping exercise was on the cards. Somebody might buy the whole and sell off a few parts. Somebody might buy the house and buildings to develop, – and let the land. Some keen rich townie might buy it all, but only want to play about with one field.

There were several possible equations, one of them could be the answer to Sep's problem. He would have to

77

keep his ear to the ground . . . grab any chance that presented itself. Willie agreed; what they needed was more land to spread the same costs over more acres. A slice of Hindhope, just over the road, would be ideal.

After supper, when he knew it would be full, Sep went down to the White Hart, not so much for a pint (though he did manage to force a couple down), but rather to test the water . . . listen to the chat. The Hart was the fountain of all local knowledge. A sophisticated gossip computer with access to a mountain of information. Feed the right programme in, press the appropriate keys, – and all sorts of astonishing half-truths could come up on the screen.

'They tell me Hindhope's been sold to a chap from County Durham with bags o' roll-over money,' said Alfie Sanderson. 'He's got t' spend it or the tax man'll have 'im. . . .'

'Aye now, I heard that,' somebody else butted in. 'Sold t' the Coal Board for open-cast. . . .' That was the programme. Up till then they had only been discussing Newcastle United and their leek trenches. Now they were buying and selling land like Texan millionaires.

Geordie Dodds said it was a moderate farm and he didn't want it at any price. He said he'd once had two bullocks stray over there and when he got them back they'd lost weight.

Gordon Rutherford who (everybody assumed) must have been left a lot of cash from somewhere because he'd bought a new tractor this year, claimed the farm wasn't all *that* bad . . . and he had heard a posse of businessmen from Leeds were interested. 'Huh,' grunted Willie Turnbull doubtfully.

Percy Elliot from the mart said he knew nothing, – but nobody had asked him, and he said it much too quickly, in a manner obviously designed to suggest he *did* know something. Chances were, therefore, he knew nothing.

Frank Blenkinsop said he'd heard Fertility Farming Enterprises were likely to have a go.

'Y' don't mean that useless co-operative outfit up

the main road, grunted brother Arthur dismissively; 'no chance . . . they'll be broke by this time next year.'

'D' y' think so?' asked Frank.

'Absolutely. They don't know what they're playin' at man, – they do everything by computer y' know . . . it *costs* everythin' t' the last penny. Have y' ever wondered why their winter wheat's lookin' so bad?' (he didn't wait for a response) 'well I'll tell y'. . . . Twice over with one of those monster discs, and the computer says that's enough, y' can't afford any more cultivations . . . sow it now! The computer's the boss, so they went ahead and drilled it . . . and a right patchy mess it is. I'm tellin y' . . . they won't last!'

Sep casually offered the postman a cigarette. 'You heard anything, Harry?' he asked.

Harry took a Silk Cut from the packet and lit it himself with one of those 'throw-away' lighters. 'Well now,' he said slowly, 'I can tell y' Les Stevenson's very keen. . . .'

'He'll just want the buildin's,' said a voice from the domino board. The man laid his double-six cross-over on the end of the line and said, 'pick the bones out o' that.' Harry drew on the cigarette. 'He might want more,' he said. 'There's talk of a golf course. . . .'

Sep didn't say anything, but he hoped for once Harry had got it wrong.

10

Down at the village farm Geordie Dodds sat on the croft gate watching his dog and the Suffolk tup eye-balling each other, six feet apart. Spot was probably thinking to himself, 'if that sheep moves, I'll nip his nose. . . .' The ram wondering, 'who the hell does that dog think he is? Come an inch closer, bonny lad, – and I'll butt y' to death. . . .'

'A useful-lookin' sheep,' Geordie reckoned, 'bold head, good bone, lengthy.' He had considered selling it as a lamb, and changed his mind at the last minute. 'One of m' better decisions,' he muttered. Well, his first offspring looked strong enough . . . to buy one that good at Kelso in September might cost £500 . . . maybe more. Y' had to consider these things. No point in spendin' money if y' didn't have to. . . .

The spring sun was on his back, no one about, only little country sounds . . . a blackbird whistling, showing off, a sheep bleating something of little consequence, miles away. Tommy Cleghorn cutting the grass in the churchyard, the strimmer whining around the gravestones. Geordie's thoughts were wandering about, not going anywhere special. He heard a heifer blaring. 'What's the matter with her?' he wondered, but the lady didn't repeat whatever it was she had said.

A wagon came slowly through the village, heading west with a load of fertiliser. Geordie turned and nodded to the driver. He didn't know the fella's name but he recognised the face. The driver remembered Geordie from the marts, and raised a hand from the wheel as he went past.

The dog and the tup had become bored with their confrontation. The tup went off to sniff at a couple of geld ewes through the fence. Spot was now lying by the gate, head on paws, brown eyes looking up at the boss thinking, 'Isn't he going to do anything today? . . . Just sitting on that gate, like a big crow. . . .'

A rabbit scuttled across the far side of the croft. Spot saw it, raised his head, but knew he mustn't give chase. Geordie saw it too. 'Bloody things are comin' back,' he mumbled to himself; 'myxie doesn't seem t' fettle them any more. . . .' The 'thing' stopped for a moment, sat up, looked about, then hopped through into Trevor Pratt's garden and disappeared. 'Not much t' eat in there,' thought Geordie, 'but you'll not be disturbed, they're never at home.'

The Pratts had bought the old ruined Smithy, with nearly two acres of land, a few years ago, knocked most of the building down and built a modern four-bedroomed house from which they commuted daily to Newcastle. Their 'field', as bare as a board, was stocked with two pot-bellied ponies and seven pedigree Jacob ewe lambs. They had at least two rabbits in the garage . . . probably a lot more by now!

Geordie remembered the blacksmith shoeing horses in that house. It wasn't all that long ago, was it? Oh yes it was! Nearly forty years. He imagined a brief whiff of the old smells. Now the place had a wrought-iron sign on the gate, 'The Forge', and Monday to Friday Mr Pratt reverses his big black car out of the drive and heads for the offices of Drew, Pratt and Illingworth, Chartered Accountants.

Quarter of an hour later Mrs P stutters out in her white oriental hatchback, and proceeds via St Mary's High School for Girls to her job as a P.P.S. at N.E.I. She tries to ignore Amanda and Tabitha bitching in the back, and doesn't put the choke in till she's onto the by-pass.

Tommy Cleghorn had moved into the vicarage garden to cut the lawns. 'Nobody there either,' thought Geordie. Certainly no vicar. When the last incumbent retired, he hadn't been replaced. Now the vicar from Nethercote shepherded four parishes, four flocks. The vast, draughty old vicarage, which every winter froze the cassocks off the clergy, had been sold for a fortune to a company director with an income big enough to install oil-fired central heating and add a conservatory. Now Mr Gordon Graham goes off to his 'soft furnishings' factory each day in the spotless 4.2 litre Jaguar, with personalised number plates . . . GG 100. Mrs G.G. is a physiotherapist at the 'General'. Geordie was surprised she still worked, but maybe they weren't all that 'flush' . . . with two lads at boarding school and another at university . . . and that barn of a building to keep warm . . . and the Jag to feed. Geordie fancied a ride in the Jag some day. He imagined himself driving very slowly through the village giving the royal wave; Spot sitting on the passenger seat, smiling . . . and slavering on the leather upholstery.

But who would they wave to? Everybody vanished during the day. The blacksmith was gone, the vicar, the policeman (gone to sit in a Panda on the A1 while vandals tore the lead from the church roof and emptied the collection box).

The two school teachers, Mr and Mrs Dawson, had moved into the village last year, and they lived in the 'Copper's House' now. They teach English and Music in the city somewhere, spend their winter holidays ski-ing in Austria, summers marinating in the Med, seldom at home. No kids, no cats, no dogs, no budgies . . . nothing to keep them here.

Their house and Burn cottage had once been 'tied' to

82

Hindhope Farm. A family of five or six, and possibly a live-in bondager as well, crammed into four rooms. Geordie remembered wee children sleeping in the bottom drawer of the kitchen press. If you lived there you worked at Hindhope, – no argument. If you fell out with the boss, you moved, – 'flitted' in May or November, your cart loaded with chairs and beds, pots and pans, wife and kids . . . only to swop jobs and cottages with some other nomadic labourer over the hill. Those two cottages were loo-less till 1958. A privy in the back yard, one shared tap that froze in winter, paraffin lamps. Rent free, potatoes free, milk and eggs free . . . maybe a pig . . and a living wage . . . just.

Geordie remembered the village school and the merry noises that erupted at playtime and the dinner break. Frantic football games in the field, – cowboys and indians, conker competitions ('mine's a twenty-niner'), girls doing handstands with skirts tucked into their knickers . . . skipping ropes. The bell ringing to summon them back to the classroom. He recalled bonny, plump young mothers collecting snotty-nosed six-year-olds in the afternoon. But the school closed in the early seventies, – Geordie couldn't remember the exact year. Some country kids had to catch Batty's bus at 7.30 in the morning now, and sit there while it meandered around half a dozen parishes, picking up at farm gates and remote road ends, until it unloaded them all at quarter to nine outside the comprehensive.

He remembered old Mr Prentice the headmaster who used to live in the school house. Nobody gave *him* any 'lip', that's for sure, no disruptive elements in *his* day. But of course y' could clout a cheeky kid then, before everybody became 'enlightened'. Geordie grunted a 'huh' at the thought, and Spot looked up to see if the boss was moving yet.

A builder from Tyneside bought the school. His lads knocked it down along with the outside toilets, fashioned a few alterations, and ended up with a good enough house, which still managed to look like a village school.

Bernard Beresford lives there now, a lecturer at the University . . . Environmental Studies . . . an expert, always making speeches about pesticides and the greenhouse effect. You could see him on the telly nearly every week, standing in the middle of a river, looking like a well-dressed poacher. Every time Geordie was spreading fertiliser he would rush across the road to mutter dire warnings of nitrates in the tap water. Chop a rotten tree down for logs, and he would accuse you of tearing up all the hedgerows. Clean out a ditch, and the man became upset about some endangered beetle. He once told the lads in the White Hart that farmers were just greedy capitalists and no longer fit to be custodians of the countryside. That got them worked up a bit.

Frankie Blenkinsop told him they'd been doin' a canny job for a few thousand years, – who did he think should take over now?

The Professor didn't answer the question, just ranted and raved about the threat to something called a silver-spotted skipper.

Frankie's brother, Arthur, said he liked to see butterflies as much as anybody else, – but didn't think there was much demand for them in Presto's.

'My point exactly,' exploded the academic. 'You're only interested in the profit motive, regardless of the long-term effect, or the suffering inflicted on your animals. . . .' He began counting off the crimes on his fingers, – 'Hens in cages, calves in boxes, pigs in sweat pens. . . .'

'Bollocks!' said Willie Turnbull (who had to be really excited before he said anything).

'It's perfectly true,' insisted Beresford. 'You've all abandoned the time-honoured principles of good husbandry for a dependence on drugs and chemicals. Money rules, OK!'

Willie gave him the sort of long, hard look he would normally reserve for a dead yow. 'You bet your bloody life it does,' he said. 'Y' don't get far without it . . . but if a farmer doesn't treat his stock properly they're not likely to prosper, and neither is he!'

That was a long speech for Willie, but he wasn't quite finished. 'M' father used t' say, "you'll never make much money feedin' cattle . . . but you'll certainly get nowt for starvin' the buggers!" ' After that, he drank up and left.

Geordie sat there on the gate thinking about Mr Beresford for a few more moments. He was lecturing in America this week, and his wife had gone with him, so that house was empty too.

'So there's nobody there either,' said Geordie out loud to himself. 'A gang with a big van could clean this village out on the right afternoon . . .' (Spot looked up again). No village bobby, no school, no vicar, no shop. . . . That's right, he thought, there used to be a shop, forgot about that. Margaret (he couldn't remember her second name), she used to run it. Pensions, family allowances, postcards, sweets, baccie . . . school kids spent their pocket money there. She was open all hours, – worked like a Ugandan Asian, and still couldn't make it pay. Anyway she packed it in long ago. Old Fred Little and his missus lived there now . . . both gettin' on a bit, retired. He had something to do with roads, maybe a surveyor. Great gardeners . . . garden always looks good, full o' fruit 'n' veg.

The rabbit (maybe a different one) scuttled out of the accountant's plot, and looked both ways before loping back down the croft. It proceeded to the bottom fence, past where the tup was chatting up his 'groupies', and turned into the lane leading away to Paddock House where Peabody the solicitor lived. Geordie watched it hop over the highway into his manicured garden. 'Wouldn't go in there,' he said. 'He'll prosecute y' for sure. . . .'

Aye, the village had certainly changed; Geordie supposed it was inevitable. But wait a minute, not all of it. . . The White Hart hadn't altered much, it was still a spit-and-sawdust place, unique really. No shiny chrome, no cracked red plastic. No point in ordering a candle-lit dinner for two with the house wine in there! However the locals liked it the way it was, – the ale was good, the darts team keen. Dryborough beer mats, Scottish and Newcastle ash trays. Customers tended to stand up, unless they were playing dominoes, – and they talked to each other. There was no muzak.

And the church probably hadn't changed since God knows when. The first one was built in eleven hundred and something (Geordie had read that somewhere). He imagined it must have been knocked about a bit since then, rebuilt, added to. Nickers's ancestors had brass plates on the walls of the knave, commemorating a captain who fell at the Somme, a diplomat buried in the Punjab. Outside, the churchyard had looked full for as long as Geordie could remember, yet they still found a space when it was needed. Tommy Cleghorn dug the graves, and nobody had offered to take the job from him. He was bound to come upon somebody's dry old bones one of these days. Dry old bones of long ago, farming folk mostly . . . and the odd aristocrat, and an occasional cleric. Grandads, mothers and bairns, whole families wiped out by a plague that had become no more than an inconvenience now, – cured by a pill.

To be fair, thought Geordie, maybe a casual traveller driving through the village might not notice the changes he saw. The houses looked smarter, more prosperous, a few extensions stone-built to match the rest of course (the planners had insisted on *that*). But everything else looked more or less as before. Geordie shifted his backside a little on the top rail of the gate. But it was different – perhaps it was the people inside the houses who had changed. Different attitudes, different priorities . . . well-heeled townies determined to be 'countrified'. And who could blame them? Certainly not Geordie. 'I don't want to live anywhere else,' he muttered. 'When I leave here it'll be feet first, that's for sure!'

But it wasn't just the village, was it? The whole countryside was changing . . . and not for the first time. For years y' couldn't produce enough, and now they say there's too much. Quick, quick, slow . . . a different tune altogether. And if the rhythm changed y' had to dance a different step, didn't y'? – or somebody stood on your toes, or you fell over, or just looked ridiculous. 'No good rockin' 'n' rollin' if the band was playin' an old-fashioned waltz, was it?' (He said the last few words loud enough for the dog to hear, and Spot looked up wondering what it was about). And then prices had gone sky high, – crackers! Even Nickers had been tempted; he'd sold the Forge, a couple of other plots, and one or two cottages. . . . Y' couldn't blame *him* either. It must cost a fortune to keep all those horses and hounds in bran and biscuits.

And now he was selling the Forsythe place. . . . Wonder who'll buy that? thought Geordie. They'll split it up, I expect, that's the way it goes now. Time was when a host of farmers' sons would be keen to have a go . . . but that was yesterday. There's a shortage of them now, so I'm told, – better educated, better prospects elsewhere, more money. (Geordie's mind was rambling on.) I'll tell y' somethin' . . . farmin's become a bit of an embarrassment, – the politicians don't quite know what to do with it. In fact y' could argue they're just payin' us peasants t' piss about until some disaster comes along, and they run out o' grub. . . . 'Isn't that right, Spot?' he said.

The dog raised his eyes again, but the head stayed down on the paws. He knew he was being consulted, but wasn't sure what the subject was.

Another wagon: 'Brownlee, Haulage, Darlington' displayed above the cab, the load covered by a heavy green tarpaulin. It turned left at the war memorial and moved slowly up the lane towards Anderson's Store.

Now if we're talkin' about changes, thought Geordie, or progress or transformations, – *there's* a business that has t' be the prizewinner. Most locals could remember when Joe Anderson only sold buckets, spades, fencing wire, your everyday pots 'n' pans, perhaps a few bags of sugar beet pulp, a mineral block, wellies, – not much else, all from a crowded wooden hut. He would never survive competition from Agrigrain or Farmco. The big anonymous companies would drive the poor chap t' the wall, or swallow 'im up . . . them with their massive investments, competitive prices and computerised accountancy. Ah, but Joe kept pluggin' away, provided what folks wanted . . . and delivered it on a Sunday night if they were desperate. And he smiled a lot, – that helped.

On reflection he must have been a pretty shrewd bloke as well. Geordie remembered when the store first began to stock electrical goods. Just plugs, switches and radios to start with, – later washing machines and fridges. His timing seemed pretty good. He expanded just as people discovered they could afford these items . . . and had to have them. So if y' needed a pot of paint to decorate the kitchen, a telly for the lounge, a seat for the garden, – y' didn't have t' travel miles . . . Joe had it. If he hadn't, he'd get it for y'.

He built a big new shed with loads of space, and a bungalow for himself. There were some folks who wondered if he'd gone too far. He went on growing though, – four vans on the road, local drivers who knew everybody and put the delivery in the right place, even when the customer wasn't there . . . no problem. Medication for your yowes, stereos for your kids, sprays for your crops, a microwave for your wife . . . light bulbs and bird seed, shoes and sealing wax, – it was all available at Joe Anderson's.

Geordie remembered the year when flea beetle attacked the turnips. They attacked everybody's turnips that year, and there was a tremendous demand for Didicom spray. He phoned Agrigrain in desperation.

'Sorry,' said the bored female voice (Geordie imagined she was painting her fingernails), 'we haven't any left. . . .'

'How about trying your other branches?' he asked.

'Sorry,' she said, 'we haven't any left anywhere. . . .'

'Can I speak to your boss?' Geordie snarled.

'Sorry,' she said, 'he's not in today. . . .'

He phoned Farmco. 'I'll put you through to our chemical department,' said the weary voice at the other end, and left him listening to The Blue Danube played on a recorder.

'Yes,' snapped the chemical department, eventually.

'I need some Didicom for m' turnips,' explained Geordie.

'You'll have to wait while I refer to our computer,' said the voice. 'This will reveal whether or not we have this product in stock. . . .'

It sounded super-efficient. Geordie waited patiently as the soulless gadget searched, and the Danube flowed again. At last chemical department came back on the line. 'Sorry,' he said (without sounding all *that* sorry), 'we're sold out . . . I believe there's a world shortage. . . .'

Geordie phoned Joe at the store across the road.

'Didicom?' said Joe thoughtfully. 'Everybody's been after that stuff . . . sold the last can half an hour ago.'

Geordie swore, 'I need it pretty badly,' he said.

'Oh don't worry, I'll get y' some,' promised Joe. 'It'll be with y' tomorrow. There's a firm in Doncaster I deal with, they'll put it on the train this afternoon.'

'Are y' sure?' asked Geordie. 'I hear there's a world shortage. . . .'

Joe phoned his pal Bob in Doncaster, who sent a lad called Tony down to the station with a one-gallon can of flea-beetle spray tucked under his arm. The package caught the 3.45 p.m. train to Newcastle, where Joe's cousin Sylvia collected it, and put it on the No. 52 bus heading north. Harry the postman picked it up from the depot at five the next morning, and delivered it to the village farm at seven o'clock. The turnips were sprayed by mid-afternoon, – and when the invoice came a few weeks later, Joe hadn't charged any extra.

Tommy Cleghorn had finished at the vicarage, and came past on his way to the Pillick residence, carrying a chainsaw.

'Full o' busy?' Geordie asked. Tommy leaned on the gate, wiped imaginary sweat from his brown brow, pushed back his red hair and proceeded to list his jobs: those already completed, those still on the agenda, and those that might never be started this side o' the New Year. . . . Geordie wished he hadn't asked.

Tommy was always run off his feet, at least if you believed all his blather. Known sometimes as the 'Clockwork Turnip', he was always on the move. He lived with Linda and their two children in the Glebe cottage, right next to the vicarage. He paid the rent religiously to the Church Commissioners, but he didn't believe in Income Tax. Seldom in one county for more than a few days, and always paid in cash, – he was a wandering nightmare as far as the Inland Revenue were concerned. Very occasionally some bureaucratic bloodhound would catch up with him. But Tommy always managed to produce a mountain of blurred invoices, illegible receipts, sick notes, and battered black notebooks of astonishing financial fiction, – all of which left HM's Tax Inspector feeling very inadequate.

Tommy put down the chainsaw to roll a cigarette.

'What are y' goin' t' chop down?' asked Geordie.

'Just trimmin' a few branches for Mrs Pillick,' he said.

'No big trees, I hope.'

They both laughed. 'No,' said Tommy. 'No big trees.'

It was about a year ago he'd felled an old sycamore up beside the garage; said he never saw the wires through the branches. When it came down it blacked out all the village and about ten farms.

On another occasion he'd been working with three other lads cutting down some timber for Sir Nigel. Tommy was well through the trunk of a very tall tree when somebody realised this one was swaying all ways in the breeze . . . nobody was sure where it would drop. Then it dawned on them that their vans and pick-ups were probably in great danger. Without saying a word, they all did a Le Mans sprint to their vehicles, started up and roared off in second gear. Two of them crashed into each other, and the tree buried all the bait bags.

But you wouldn't hear a wrong word about Tommy, – a country character in a world where characters were becoming an endangered species. He would tackle any job, pass the time of day with bin men and baronets, and in spite of the odd disaster, never fell out with anybody. Well, hardly ever. There *was* that story of the milkman who tried to charm the nightie off Linda every morning, while Tommy was away working over beside Carlisle. Somehow our hero got to know about this, and went straight round to the Dairy and confronted the bloke. The milkman recognised him immediately. 'I should warn you,' he said menacingly, 'I'm a black belt at kara—'

It was at that moment that Tommy's 'Rogerson of Rothbury' toecapped boot came up and clouted him in the groin. 'Bugger yer belt,' said Tommy, and left the man groaning among the empties, clutching his discomfort.

Geordie eased himself down from the gate. 'Well,' he said, 'we'd better get on, – big day tomorrow. . . .'

'The sale?' said Tommy, placing half a cigarette behind his ear.

'Aye, the sale.' Spot was prancing about, eager to be doing something at last . . . setting off for a few yards, coming back to the boss again, agitated. If he could have spoken, he would have said, 'C'mon Geordie, hurry up, – the tup's missing!'

'Where's that damned tup?' Geordie asked, looking round the croft.

'He's over the fence,' said Tommy, 'tryin' t' seduce Mrs Pratt's black 'n' white Jacobs. . . .'

'It's a bit late,' muttered Geordie, 'He did the job last November . . . but Mrs P. doesn't know that yet. . . .'

11

Farm sales were rather sad affairs, Sep thought. A lifetime's flotsam laid out in a field for the community to pick at. The farmer playing his last card, everybody watching . . . no secrets.

The lads needed a good sale, and although it had been extensively advertised, Samuel Forsythe was seen at dawn nailing signs on trees and telegraph poles for miles around. *To the Sale*, – arrows guiding the chequebooks to Hindhope.

As Sep drove in half an hour early at one-thirty, the car park field was already filling up. So was the beer tent. 'What y' after today, then?' somebody would ask, and a man who was keen to buy maybe a dozen young bullocks would smile and say, 'Oh nothin' really, – just came for a look. . . .'

Old friends, who met every week at the marts, eyed each other like opposing centre-forwards before a cup final. 'We'll get nowt cheap t'day,' one of them would say, 'Sep Robson's here.' And Sep (who was meant to catch the comment) looks round, laughs, and says, 'Oh, it's you

Charlie. Are y' never at home? Do y' do no work at all these days?'

A big man with a Yorkshire accent and waterproof leggings up to his armpits leaned against the cartshed wall, whispering into a black cordless phone. He looked furtively about, turning his back when anyone passed by, like a conspirator, a foreign agent plotting the overthrow of a local regime. 'Aye, they look not s' bad,' he muttered, 'depends on t' trade. . . .' He listened for a while, said something about 'traffic on t' A1', mentioned 'a filly 'runnin' in t' three o'clock at York', – and finally pushed in the aerial, straightened himself up, and waddled off towards the sheep pens.

Now his clandestine call was finished, he took on a completely different persona, greeting all-comers as close friends: 'Now then young man, are y' fit?' – and before the 'young man' (who might well be in his late seventies) could respond to this unexpected greeting, the Yorkshire spy had moved on.

He came upon Geordie drawing out some warranted ewes with one red dot on their backs from a group of 'udders only' with two dots. 'Now then, young man, what 'ave we got 'ere?' he asked, putting his hand round Geordie's shoulder.

'Some right good yowes 'n' lambs,' grunted Geordie, looking pointedly at the man's hand. He had no time for any interruptions, and certainly didn't approve of such familiarity, – especially from a bloke who had sneaked up from south of Darlington. But Geordie knew well enough who the man was, and so did the auctioneer. If Ned Forster from Thirsk was here, it wasn't just for a day out in the country.

Ten minutes to kick-off, – cars, pick-ups and wagons still coming into the field. It looked like being a good crowd.

Sep and other 'experts' were viewing the cattle. The bewildered beasts had been drawn into small matching lots, penned in the hemmel and loose boxes around the yard . . . ready to run directly into the ring. Some of them had been parted from their mates, and were complaining bitterly. 'That 'n' shouldn't be in there,' someone said, pointing to a bullock with too much gut; 'spoils them.'

'They'll shift,' declared another connoisseur, peering into the stable at a bunch of lean stirks with a lot of hair, knee deep in clean straw.

'Might not look s' big when they come out,' warned his companion.

Thomas and Samuel were very nervous. They had been up at the farm since before daylight, checking the various lots for the umpteenth time. They desperately wanted the sale to go smoothly, no cock-ups, no embarrassments. It was all too theatrical for them; everything they possessed was up on the stage, the audience taking their seats, – the critics sharpening their pencils. Thomas was sitting on a straw bale at the sheep pens, waiting, idly scratching Moss's left lug.

'I wish it was tomorrow,' he said quietly, apparently to the dog . . . and a lady in a Barbour coat, green wellies and yellow bobble hat looked down at him with very surprised eyes.

Arthur began selling the 'implements' twenty minutes late. Samuel had lined them up in two rows, lots 1–48, in what he considered ascending order of value. A collection of chipped earthenware troughs first, the tractors last. There were a dozen troughs, and Samuel wished there had been a hundred. After some frantic bidding, mostly from well-dressed people he had never seen before, Fred Little from the old shop bought them for his garden. 'They'll be full of geraniums next week,' he said happily. 'Y' can't get proper troughs like this any more, – much better than those awful plastic tubs at the garden centre. . . .'

The crowd followed Arthur along the line, Percy the clerk recording names and prices. Well-worn spades that had dug holes for straining posts and dead sheep; graips and forks that tossed muck and hay, – they all got a bid.

95

Past the wheelbarrow, the harrows, the link box, on to a set of discs, gang rollers and the fertiliser spinner. Considering everything was aged, the bidding was frisky. The little tractor made £250 more than Samuel expected. The Ford 4000 (with a slow puncture in a front tyre, which thankfully stayed up) was sold for twice the guesstimate.

Everything found a buyer, and the pieces of surplus furniture from the farmhouse, the items that wouldn't fit into the cottage, went like a bomb. A massive pine dresser, a mahogany wardrobe that had to be dismantled to get it downstairs, a marble wash stand, the dining room table with elephantine legs and cup marks on the polished top. The sofa from the kitchen, tufts of horse hair peeping out and smelling of dog; horse harness from the stable, – all bought by an antique dealer from Newcastle, who wouldn't be denied, even though Arthur ran him with a few phantom bids. The only item that failed to attract any response was the grain crusher, still fixed to the concrete in the meal store, and there it stayed.

'Ah well,' smiled the auctioneer, 'y' can't win 'em all,' – and like a pied piper he led the gathering over to the sheep pens, where Thomas was anxiously waiting with the first lot already in the ring.

The sheep were drawn into age groups, beginning with some canny three-crop mules and twins, and ending with a dozen very old Blackies and singles, the superannuated tups and the hoggs. Cordless phone bought most of them, but he didn't have it all his own way. Willie Turnbull got some for his grass park, and Sep topped up his numbers with a pen of four-crops with strong lambs.

The last sheep into the ring were a couple of rough mule hoggs with horns who had survived as best they could after their mother expired last summer. The lady with the Barbour bought them, and appeared to be overjoyed. 'We'll have little lambs of our own next year,' she told her spotty daughter. 'We can borrow Mr Dodd's ram. . . .' Thomas reckoned they might be disappointed. The ram certainly would be.

96

After that Arthur sold the pens, and moved on to the cattle. The crowd followed him to the yard, and climbed onto bales set out around the walls. Samuel was in charge here. He brought the beasts in, Thomas moved them about the ring and Geordie, with two lads from the mart company, put them back where they had come from, hopefully without any mix-up. Arthur, perched on a forty-gallon oil drum, was eager to proceed. Black clouds were building up in the west, a few spots of rain falling. Buyers would want to be loaded and away before dark. Some sheep and most of the implements were moving already.

He didn't need to worry, – the cattle were 'on fire'. Arthur rattled through them in forty-five minutes of keen bidding. It might have been even quicker but for a cross-Charolais stirk who came out of a dark loosebox like a Spanish bull with an Exocet under his tail, and roared around the yard looking for a way out. Several farmers waved their sticks and shouted 'whoa!' at him, but that didn't work and, as he charged the crowd at the bottom end for the second time, they wisely parted like the Red Sea and let him go. Arthur had Sep in at £305 at the last sighting, and knocked it down quickly. Sep said he didn't think it was too dear, – provided somebody caught 'im before the bugger got to the western bypass.

And so it ended. The hammer fell for the last time; Arthur thanked his audience, even though a lot of them were already half-way to the car park, or settling their bills. Ruby was serving tea and biscuits in the kitchen for the last time.

Outside, 'cordless phone' was giving instructions to his wagon driver. Barbour coat was pushing her two mule wethers into the back seat of a Volvo, while spotty daughter fed them digestives. They had names already. Alfie Sanderson was looking at some lean bullocks he had bought, telling anybody who might listen he would have them fat by July. Fred Little was quietly loading earthenware troughs into his hatchback. He would have to make three journeys: 'hadn't realised how heavy they were'. The last of the cars rolled gently over the rigs and out of the field. Wagons were lined up waiting their turn to load. It was raining steadily . . . a good April rain.

Thomas and Samuel opened a bottle of Grouse, washed out some tea cups, and shared it with the auctioneer. 'It's been a good sale,' declared Arthur. 'I think you'll be pleasantly surprised when you see the figures. . . .'

When everybody else had gone the two brothers had a last look round. Thomas thought it looked like a battle-field after the armies had withdrawn, when the guns were silent, the dead buried. Out in the sale field only a few bits of equipment remained, the reaper, some posts and rails, two hay hecks . . . and the Ford 4000, with a flat front tyre.

'Well, that's it,' smiled Samuel. 'No more lambin's, eh? Retired!'

'I think I'm lookin' forward to it,' said Thomas. 'I'm knackered.'

So Ruby drove them back in the pick-up. The cottage was home now. After supper they wandered down to the White Hart and bought everybody a drink. Samuel didn't think he could sleep without a pint or two on board.

At Clartiehole Sep was watching the nine o'clock news when the phone rang. It was Alfie Sanderson. 'I've got your mad bullock,' he said. 'He's barred up in the byre here. If y' come over on the tractor in the mornin', I'll lend y' my horsebox to take him home. . . .'

'Thanks,' said Sep. 'Is he all right?'

'Yes, he's all right,' Alfie replied. 'But when I got a proper look at 'im, he's a lot smaller than I thought. I think y' might be disappointed. . . . See y' tomorrow.'

12

Spratt and Mackeral printed over a hundred glossy Hindhope brochures. They posted fifty to assorted clients from the Highlands to Hampshire, and received a further thirty enquiries by post and phone. Yet, two weeks before the closing date, there were no firm offers in. Timothy Williams-Wetherby from the Harrogate office appeared undismayed.

'Fear not, Sir Nigel,' he said cheerfully. 'These things take time, you know. A prospective purchaser will have to put his own house in order, make funds available, that sort of thing. . . .' He crossed one long, cavalry-twilled leg over the other, exposing bright yellow socks. 'We're talking real money here, of course,' he declared.

'What!' roared Nickers. 'On the bloody contrary, we're not talking money at all, – real or imaginary. All I've heard from you Johnnies is a lot of hypothetical clap-trap. Can't you tell us who's likely to buy the damned place . . . what?'

Newcastle-Browne was staring out of the long French windows, over the gardens to the parkland beyond. 'Perhaps you could tell Sir Nigel who the likely candidates might be,' he said without turning round. 'I mean, who has actually shown some real interest, who's inspected the property. . . .'

'Bollocks!' said Nickers angrily. 'I don't care who buys the farm as long as he pays over the odds, what! We need the money, and pretty damn quick. . . .'

'Well, now, gentlemen,' said Williams-Wetherby (there was another flash of yellow), 'I believe we can boil it down to four or five contenders, with what we might call the serious money.' He leaned back, and with his right index finger on the little digit of this left hand, began to list the probable runners.

'There's this fellow, Sanderson,' he said. 'Local chappie, knows the farm well, shown a lot of interest, walks over the land at least once a week . . . talks a lot . . . maybe all bravado of course, – can't be sure at this stage.'

He moved to another finger. 'Lord Richards of Bradford might have a go—'

'What!' Nickers was on his feet. 'You can't mean "Filthy Richards", chairman of Bradford Bricks? I was at Charterhouse with the smelly little creep . . . thick as a plank!'

'The very same,' smiled Williams-Wetherby. 'Charming man, he's been buying land all over the county recently . . . has aspirations to be a major landowner, I believe . . . create a new dynasty perhaps. His staff have certainly inspected the farm. . . .'

'Bloody cheek,' snarled Nickers. 'The man's got no breeding at all, can't talk properly . . . has one of those trouble at 'mill accents . . . decidedly common. Daphne says he nipped her bottom at the Conservative Fête!'

'He's very well orf,' said Newcastle-Browne, anxious to be involved somehow. 'Owns 60 per cent of the company, and the shares are doing well . . . they say he could be worth twenty million at least, – maybe more!'

'And his wife's a tart,' said Nickers. 'We can't have people like the Bradfords owning half the county . . . no bloody clarse. They strike it lucky, make a bob or two, and before you know it, they've bought an estate and are playing Lord of the bloody manor. . . . What!'

'Rather like your great grandfather, George,' said Williams-Wetherby, and immediately wished he hadn't. He moved quickly onto another finger. 'Stevenson,' he said, 'Leslie Stevenson, builder-developer, local boy made good, bit of an entrepreneur, the sort Mrs Thatcher delights in. . . .'

'Why on earth would *he* want to be a farmer?' asked Nickers.

'Oh, I don't suppose he does,' said W.W. 'I suspect he'd re-shape the bricks and mortar, dispose of the land, and move on with a nice profit. He's done it before, apparently. . . .'

'Good God!' Nickers needed a drink. He moved to the sideboard and poured himself a large Glenfiddich. He didn't offer it to anyone else. 'So far,' he said, 'we've got a windbag who probably can't afford to buy, a nouveau-riche rascal with delusions of Dukedom and a bloody asset-stripper who'll simply dismantle the place!' He drained his glass. 'Haven't you got a proper peasant on your list?' he asked loudly.

'Well yes, I think we have, – possibly two. . . .' Williams-Wetherby had reached his left index finger. 'There's a chap from County Durham with oodles of open-cast money. . . .'

'Open-cast money?'

'Yes, his farm has been erased by the Coal Board,' he said. 'Everything bulldozed, dug up . . . a moonscape with a ruddy great crater in the middle. However, he was very handsomely compensated, of course, and now he'll be very handsomely taxed if he doesn't spend it. He's called Sedgefield, nice fellow. He's looked at several farms already, but I think he'll definitely offer for Hindhope. . . .'

'Does he hunt?' asked Newcastle-Browne.

Williams-Wetherby either didn't hear or chose to ignore the question. 'And we have another farmer from Cumbria with lashings of quota cash,' he said.

'Quota cash!' Nickers was exasperated. 'What the hell is quota cash?'

The man from Spratt and Mackeral smiled the lofty smile he used when explaining simple things to simple souls, like a politician on TV. He leaned back in his chair and stretched his legs, the trousers creeping up to the top of his socks.

'Dairy farmer,' he said. 'Cyril Brampton, family in cows for generations. Three brothers in two farms with Friesian herds, very efficient . . . milk sloshing about all over the place. One of them, totally disillusioned with cows, decides he wants to grow wheat instead . . . sells a hundred cows at just the right time, and leases his milk quota to somebody else. Half a million litres at 6p a litre is a very nice little earner . . . so the man's looking for more land for corn and sheep. Much more attractive than milking cows twice a day, seven days a week, he reckons.'

Williams-Wetherby looked towards Sir Nigel and Newcastle-Browne to see if his story had registered. He couldn't be sure.

'Anyway,' he said, 'our man is really quite well orf, and can certainly buy the farm . . . there's little doubt of that!'

'Well the quicker the better!' snorted Nickers. 'I owe the bank a bloody fortune, way over the top, getting bigger every day, – and they're charging me 18 per cent. Got to get this debt thingie down or we're knackered!' He was up from the table and stomping about the room on a march that took him to the Adam fireplace (where he paused to warm his backside), then to the window (where he peered out at the Capability Brown landscape), back to the Chippendale table (to pick up his empty glass) and on to the Mousey Thompson sideboard for a refill. 'Anyone else?' he asked, holding up the bottle.

Newcastle-Browne briefly considered saying, 'I thought you'd never ask,' but thought better of it.

Williams-Wetherby said, 'Lovely. With a little soda please.'

'So is anyone else likely to have a go?' Nickers wanted to know; 'is there a dark horse lurking somewhere, what?'

'Well one can never be sure in these matters,' said Williams-Wetherby. He enjoyed being centre stage on such occasions, – especially when the set was a stately home, and his commission a major part of the plot. He knew full well that in any other cirumstance Sir Nigel would hardly give him the time of day . . . would probably gallop over him, – but not this time. 'The sale *has* generated considerable interest,' he said. 'As well as those already mentioned, we've had lots of requests for particulars, and I understand there's been someone looking over the farm almost every day.'

He sipped his whisky. 'Of course they won't all be serious, you understand, but one can never tell. I remember a sale in Lancashire last year, when out of the blue—'

Nickers chopped him off by the yellow stocking tops. 'Who else?' he demanded.

'Oh well, let me see.' (W.W. had run out of fingers and changed hands.) 'There are one or two neighbouring farmers: a man called Robson, Septimus Robson, I recall; and another fellow with four enormous sons who look like the Pontypool front row,' he grinned. 'All eager to follow in Daddy's footsteps, apparently.'

'There are only three in the the front row,' said Nickers irritably, but yellow socks carried on undaunted.

'A couple of yuppies, attracted by the good, green lifestyle, I suppose, – prosperous broker types, with "country" wives. And there's a pension fund. . . .'

'A pension fund?' asked Newcastle-Browne. 'What pension fund?'

'Oh, some pharmaceutical conglomerate, I think,' said W.W. 'Amalgamated Bicarbonate, or somebody like that. But I wouldn't take them seriously . . . these people seldom know what to do with their money, extremely fickle. One minute they're buying land all over the country . . . long-term investment they'll tell you, enormous tax advantages, lease-back schemes for the tenants, and so on. Then after a couple of years, in comes a bright new manager who decides land's no good after all, and moves everything to Hong Kong or wherever . . . better returns and all that. . . .'

Lady Daphne entered, bearing coffee in mugs. 'Coffee anyone?' she chirped, smiling her 'lesser mortals' smile at Williams-Wetherby. 'Splendid,' said Williams-Wetherby, rising to his canary yellow feet. 'Love one.'

'Have we found a nice rich buyer, darling?' she asked, her mind and her Barclaycard already on the train to Harrods. 'An Arab potentate perhaps?' she laughed. 'Someone amusing, I hope. . . .' Exactly what the lady meant by 'amusing' was not clear. It was unlikely to be anyone who told dirty jokes with a Geordie accent, or wore tartan trousers and drank McEwans best Scotch. Heaven forbid it should be anyone from north of the border. Ever since the telly showed an army of Scottish football supporters being sick in Sardinia she'd been a firm advocate of devolution.

And certainly not a socialist. She refused to contemplate the prospect of one of those awful Labour people living nearby. Daphne had once been obliged to sit next to John Prescott at a function in the Civic Centre. The man had totally ignored her, talking about the channel tunnel to the Mayor all evening. She suspected he had never been on a horse in his life! Why hadn't they invited Heseltine. . . ?

'You must appoint someone interesting, darling,' she said. 'Someone like us . . . and of course they *will* hunt. . . .'

'It's hardly a question of appointment, my dear,' said Nickers quietly. 'I imagine we'll take the best offer we can get, what!' Williams-Wetherby, who had continued to stand, smiling charmingly in the presence of Lady Daphne, said, 'W'll do our best, Madam.'

She caught a glimpse of the man's ridiculous socks, and was quite appalled.

The three men waited until she'd gone before they moved or said anything more. The man from Spratt and Mackeral sat down. Sir Nigel threw some logs on the fire. Newcastle-Browne said, 'What happens if we don't get any offers at all?'

'What!' roared Nickers, 'the idea's bloody preposterous, isn't it?' Then with a little less conviction . . . 'isn't it?'

'Oh, absolutely,' said Williams-Wetherby. 'Someone will certainly offer . . . they always tend to leave it until the last minute. I expect the brown envelopes will begin arriving at the weekend. Remember, they have till noon the following Friday, that's the deadline, – that's when we open the box, as it were,' he smiled.

Nickers wasn't smiling. 'I need about £350,000 rather badly,' he said.

13

Even at lambing time, Alfie Sanderson claimed he had no losses. Nobody believed him of course, even when (as if to inject just a smidgin' of credibility into the claim) he might add, 'well to be fair, we *did* lose a couple of old yowes, – but the stupid sods strayed onto the line!'

Ah yes, the 'great railway disaster'. Everyone except B.R. seemed aware of Alfie's little fiddle. Some had even seen him dragging a dead sheep onto the track in the middle of the night. Next day he would claim it had been struck down by a flying Intercity, and seek suitable compensation. After all, it was B.R.'s fence through which the unfortunate creature had crawled, or B.R.'s gate left open by some careless rambler. Over the years an innocent clerk in Regional Claims had reimbursed Alfie for several top gimmers (seven-crop, udders only), a potential Smithfield Supreme Champion (pot-bellied Friesian stirk with husk) and even a stack of sweet-smelling horse hay valued at a fiver a bale . . . sadly ignited by a spark from a passing Pullman. In fact the locals knew the heap had been steaming ominously for a fortnight, and looked like a grey blancmange.

British Rail eventually twigged when Alfie said he'd found a decapitated tup on the line. 'Worth a fortune,' he declared.

'When did this tragedy occur?' asked the clerk.

'Just last night,' he sobbed, '. . . in the prime of life.'

'This is remarkable,' said the voice of B.R. 'We closed the line three weeks ago.'

'Never!' exclaimed Alfie increduously. 'Come to think of it, I thought the rails were a little rusty . . . there must be a wild Rottweiler on the loose, then. . . .'

'Or even a lying toad,' said the clerk just before she put the phone down.

Alfie had walked over Hindhope six times, making sure Spratt and Mackeral were informed of his intentions . . . as well as the postman, the Forsythe boys, Percy Elliot at the mart, and as many of the general public as he could persuade to listen. He would leave his Mitsubishi Shogun parked by the road where everyone could see it. On three occasions he saw other men walking the fields or inspecting the buildings, while their wives 'redesigned' the farmhouse. One day a Range Rover was parked in the yard. Another day he saw a Volvo. Last week a Rolls, complete with uniformed chauffeur. 'Who the hell was that?' On three occasions he saw no one, and had to assume no one had seen him. It was very disappointing.

In the White Hart, a week before the final offers were due, he told a group of disinterested drinkers, more concerned with the Leek Club Raffle, that he was having consultations with his accountants and bankers the next day, prior to delivering his bid.

'What y' think it's worth then?' asked Geordie Dodds. He was going for double-fourteen to win the game and 50p.

Alfie puffed out his belly. 'Well over quarter of a million, I suppose,' he said . . . as if he considered that a trifling sum.

Willie Turnbull muttered, 'm'be more . . .' and tried to

imagine what the figure looked like when written down.

'No, y'll not get it for that,' said Geordie.

Five minutes later, standing next to Sep in the gents across the yard, he said, 'D' y' think Alfie's in with a shout?'

Sep looked up. 'You're jokin',' he said. 'The man's full o' wind, and the stuff that's runnin' down this drain here. I doubt if he'll even put an offer in. . . .'

'Well now,' said Geordie, 'he says he's meetin' his accountant and bank manager tomorrow. . . .'

Sep cleared his throat and zipped up his fly. 'Aye well, he would say that, wouldn't he. It costs 'im nowt. . . .' He lit a cigarette. 'I bet y' a quid there's no sealed envelope from him next Friday!'

'What about yerself?' Geordie asked. 'You've been sniffin' about for more land, ever since your lad began to show a bit eye, haven't y'?'

Sep grinned. 'Out of my league,' he said. 'This one's for the big boys, I reckon. . . . They say Lord Richards has been up a couple o' times. He's been buyin' everything lately . . . he'll take a bit knockin' back, I expect. . . .'

They sat on the stone wall in the car park, cigarettes glowing in the dark. It was a good, quiet spring night. 'Harry the postman says Sedgefield's been there,' Geordie said. 'He'll have a bit cash likely . . . and another fella with a 'yan tan tethera' accent. . . .'

'Harry'll know,' said Sep. 'Harry knows who's been to the farm, even if they've just been poachin'.' He eased himself down from the wall. 'One for the road?' he asked.

'Aye, dare say,' said Geordie; 'w' might just be able t' manage one more. . . .'

Next morning, tucked away at the bottom of a financial page in the *Telegraph* was a brief report that Lord Richards had sold one million shares in Bradford Brick at 200p each. 'The reason for the sale,' the noble Lord assured the market, 'is to raise funds for the purchase of land. It is a purely private venture,' he said, 'and in no way will it inhibit the continuing expansion of Bradford Brick as a major force in the construction industry. . . .'

By the time this announcement appeared in the press the shares had already fallen 15p, and by the close they were down a further ten. There were at least two investors who took a cautious but self-indulgent delight in the decline, having sold their modest holdings a few days earlier . . . for the same reason as 'Filthy Richards'. One was Les Stevenson, – the other was a broker living in Acacia Avenue.

Les had done his homework thoroughly. He had wined and dined Charlie Woodruffe at the best restaurants in town, lost a few strategic games of snooker, repaired Mrs Woodruffe's jacuzzi, and secured a promise that outline planning on the Hindhope farm buildings was a mere formality. A plot of land he'd been 'sitting on' for two years was sold to a major supermarket, eager to expand their car park, and any further financing facilities that might be required had been arranged with Thompson at the bank. His record was sound, there were no problems. If his plan worked out, borrowing could be kept to manageable levels, and paid off within about eighteen months.

Les, who knew his limitations, commissioned William Ackeroyd, an estate agent and valuer from the West Riding, to come north, heavily disguised as a bird-watcher and sworn to secrecy, to value the farm as an agricultural holding. His comprehensive report also considered development potential, the worth of the farm-house by itself, even a possible set-aside figure for the arable land. A firm of solicitors in Newcastle were instructed to perform the usual search on the property. Architects' plans for converting byres to bedrooms and loose boxes to bathrooms were already drawn.

Four days before offers closed, Les delivered his bid by hand to the offices of Spratt and Mackeral. It was for the farmhouse, the steading, and all the land to the north of the road.

*　　　*　　　*

Gerald and Prudence were never as well organised as Les Stevenson. In fact Gerald was far from convinced that their proposed bid was at all sensible. He was even hoping (sometimes, at least) that it would not be accepted. Only the persistent Prudence with her endless lamentations on sordid, polluted, cardiac-inducing city life, and her clean fresh, carefree rural alternative, had finally broken his resistance. In the end it came tumbling down like the Berlin Wall. The first stones to topple were pushed by Julian and Emma. Prudence had been to work on them, and half-term became an orgy of farmyard imaginings: calves drinking from buckets, lambs sucking from bottles, chickens scratching, dogs barking, hamsters fornicating, and ponies galloping over meadows. The pressure became intense, and often unfair: 'Darling, we simply must, – if only for the sake of the children. . . .'

Away from Acacia Avenue things were crumbling too. Other than the privatisation issues, and the on-going management of some old-established portfolios, the firm was doing little business these days. The collapse in October '87 had frightened more than widows and orphans. There were ominous rumblings of redundancies and early retirements. Old Colonel Pickering, a senior partner, didn't wait to be pushed; he took his pension and headed for the golf club. Johnnie Waterhouse, who'd only been with the firm for a few months, was offered a London job in commodities, and wasn't replaced. Martin Fraser, the whiz kid of the Trading Room, left under a cloud . . . something to do with misappropriation of clients' funds. He seemed undismayed as he drove off in his Porsche 911.

Tony Gardner, the guru of the Unit Trust section, had forecast a bonanza in the Japanese market, only to watch it plummet several thousand points. Dismayed investors, who yesterday considered him a genius, phoned to tell him he was now a 'crooked git'.

'Hang on,' he advised, 'this is but a temporary setback.' Further plummetting produced more calls, inviting him to commit hara-kiri. 'Trust me,' he said, 'this is but a technical adjustment.' However, as the market continued to head south, he received threats and even occasional visits from the 'disenchanted'. Nice old ladies were anxious to tear his throat out. The board felt he had to go; the firm's reputation was damaged, Tony had to carry the can. They filled the can with a very generous severance settlement, – and he left to take up a similar position in the City, at a much higher salary.

Gerald himself was threatened with litigation after complimenting the voluptuous Ms Hartstoppe from the typing pool on her colourful new dress. 'Sexual harassment!' she screamed, and claimed he was leering at her bosom and constantly dropping paper clips on the floor in order to look at her legs. Naturally Gerald stuttered vehement denials, apologised profusely for any embarrassment he may have caused while on his hands and knees . . . and promised never to comment on any item of her apparel ever again.

When he told Prudence about it, she called the woman 'a stupid cow', and said such incidents simply did not occur in the countryside.

Gerald had the farm valued by a firm of estate agents with a familiar name, picked from *Horse and Hound*. They duly sent him a five-page dossier on their conclusions, and a one-page bill for their bother. He found both alarming.

They visited Hindhope three times, once in the company of a young man from the valuers, who looked so young Gerald wondered if he might still be at school. However, he seemed to talk knowledgeably about the

111

EEC, MAFF, LFAs, SSIs and ESAs (whatever they were). He twittered on about woodland management, sporting rights, and a whole lot of stuff that left Gerald utterly bewildered.

The second trip was with their bank manager on the morning after the valuation arrived by registered post. Surprisingly the man from the bank appeared completely at ease with the figures and promised whatever assistance might be needed. He drew Gerald's attention to the rising value of his own property in Acacia Avenue, – and compared it with the probable price of the farm. 'Look at it this way,' he said. 'What can you buy in the housing market these days for £300,000?' No one offered an answer, so he went on, '. . . well, a very nice detached four-bed villa perhaps, with double garage and quarter of an acre of garden . . . that's all.' He paused and looked around. 'In the agricultural world you could pick up an enormous house and maybe two hundred acres for the same money. There's no comparison . . . and you'll get an income from one and nothing but bills from the other. . . .'

Gerald nodded. 'I suppose so,' he mumbled doubtfully. He hadn't bought the damned farm yet, – and he was getting bills already. The banker shook his hand. 'We'll be delighted to offer you whatever loan you may require,' he smiled. 'The security's first class, just tell us what you need. . . .'

The third safari was with the children, who ran about like young stirks let out of a hemmel on the first day of spring. They played cowboys and Indians in and out of the byres, they rolled on the grass, climbed on bales in the hayshed, – and assured their parents that if they failed to buy the farm they would kill themselves!

But Gerald havered right up to the last few days. This was a major step into the unknown. Moving to a house in the country was one thing, playing at peasantry was something else. It was like buying a massive block of equity in a company you had never heard of, – just because someone in the pub said it was a good bet. Did farmers never go bust? He remembered a quotation he'd seen somewhere. How did it go? . . . Something like, 'neither good farmers nor bad farmers ever get their just reward. . . .' Was that reassuring? He wasn't sure.

And they would need a big bridging loan. Well, they could hardly sell Acacia Avenue until they were into Hindhope. And that farmhouse! – God it was an ancient monument . . . it would cost a fortune to make it barely habitable. Prudence would go berserk in there. He saw a brief vision of a cash register ringing up the equivalent of the Venezuelan national debt on his Barclaycard.

'It's quite beautiful, darling, isn't it?' she murmured softly. She had the glossy brochure, and came to purr on the arm of his chair, stroking the back of his neck. 'I think we'll all be very happy there.'

He'd been doing sums on sheets of foolscap, calculator at his elbow . . . and he had made the decision. It was clearer now, and he was not going to be diverted.

'I'm staying with the firm,' he said, wishing he'd coughed first, because it came out in a broken soprano. She said nothing.

'The way things are going, I could be on the board soon,' he went on. 'Not many senior blokes left; I could even be chairman in a year or two, if all goes well, I think. . . . Can't afford to abandon that chance. . . . And let's face it, darling, I'm never going to be a real farmer, not in a thousand years. . . .' She was fiddling with the calculator, pressing random numbers, staring at them.

'However,' he paused and looked up her. 'However, we agree the house could be fantastic. It's only an extra twenty minutes' commuting time, and it'll be tremendous for the children . . . and for you darling, – and for me, of course,' he added hurriedly. She still said nothing, but had stopped playing with the calculator. 'We'll make an offer for Lot 3,' he said calmly, 'the house, and that little field. . . .' He stopped, waiting for a reaction.

She bent over, kissed him on the nose and said, 'Super, darling, – I'll freshen up your gin. . . .'

As she turned towards the drinks cabinet, Prudence picked up the local *Gazette* and handed it to him. There on the livestock page ringed in red ink: 'For sale, Welsh cob 12 hands, quiet in traffic, £750 to good home, telephone. . . .' Further down, another red circle: 'Pedigree British Alpine Kid, female £150 to good home. . . .' And another: 'Point-of-lay pullets for sale, £3.50 each'. More: a Blackface ewe, a pair of rabbits, another horse, an Ayrshire heifer. . . . Gerald saw them all ringed in alarming, overdraft red.

'Your drink, darling,' she smiled.

14

The morning of fateful Friday May 13th was distinctly 'Novemberish'. An unkind, drizzly east wind disturbed the trees, crept through careless open doors and over upturned collars. Cattle, recently out to grass, grazed in tight groups, backs up, bums to the weather. Dogs shook themselves, tups shelved any out-of-season erotic notions, birds didn't fly far, and some farmhouse fires were lit. Sheep nibbled on relentlessly. Spring was away for the day . . . not answering. It sometimes happens in May.

Up at the Hall the Honourable Nigel Nicholas wore an anxious look. This was decision day. The day sealed offers for Hindhope would be opened. Eleven o'clock was the deadline.

Lady Daphne served coffee in the sitting room: silver tray, best china, tongs in the sugar bowl, cream in the jug, miscellaneous biscuits. She fluttered about like a spuggie on amphetamines: 'One lump or two?' . . . 'Black or white?' . . . 'Oh, I've forgotten the spoons.' . . . 'Are you all warm enough?' She threw two logs into the grate and, still fluttering and twittering, withdrew at speed, closing the door behind her. This was a job for the gentlemen.

'Right, let's get on with it.' Nickers was becoming agitated.

'It's only five to eleven,' said Newcastle-Browne, looking earnestly at his watch.

'What!' snorted Sir Nigel. 'Near enough, – open the bloody envelopes!'

Williams-Wetherby swallowed his coffee hurriedly, spluttered a little, and opened his briefcase. He carefully withdrew a calculator, together with three pads, three pens and a paper knife, which he laid on the mahogany table like place settings for lunch. He seated himself at the top (he felt this was *his* show), Nickers to the right, Newcastle-Browne to the left. Waiting until the other two had settled, he then produced a bunch of long brown envelopes, all addressed to Spratt and Mackeral, and marked *Hindhope Sale.*

'How many?' demanded Nickers.

'Nine altogether, Sir,' came the reply.

'Is that all? What, – I was led to believe thousands had been wandering about the farm, loads of 'em . . . where the hell have they all gorn?'

'Ah well,' said Williams-Wetherby in his I've-seen-it-all-before voice, '. . . when it comes down to the nitty-gritty, they don't all come up with the money, you know. They have second thoughts, the wife doesn't want to live there, they find a better farm, after all, there are others on the market, – all kinds of reasons. I remember a chappie who roared up in his Rolls only minutes before we opened the offers, and. . . .'

'Get on with it!' said Nickers. 'Open the bloody things, – let's see what we've got, what! It's gorn eleven.'

Williams-Wetherby made quite a drama of drawing the first envelope from the pile. He opened it carefully with the knife and withdrew the contents. 'From Lord Richards of Bradford,' he announced. 'His Lordship offers the sum of £330,000 for Lot 1; that's for the . . . the entire holding. . . .'

'Ah, good old "Filthy",' said Nickers. 'Let me see it.

116

Seems a very reasonable offer, what?'

'Very fair,' said Newcastle-Browne, 'very fair indeed.' He felt he had to say something meaningful, so he added, 'but early days, Sir Nigel, – eight more to go. . . .'

The offering was passed around the trio, details written down on their pads. . . . The second envelope was opened.

'Mr Cyril Brampton from Cumbria bids for Lot 2, which you will recall is the land only, all 245 acres (he pressed calculator buttons). He offers £200,000, and that's about £816 an acre.'

The document proceeded via Nickers to Newcastle-Browne, and back to Williams-Wetherby, before he spoke again. 'Leaves us with the farmhouse, buildings and the croft,' he said, 'and I expect there'll be an offer for those. You will note, gentlemen, that Mr Brampton intends to erect sheep- and cattle-handling pens on the west side of Rabbit Lonnen, and will apply for planning permission to build a shed.' He waited for any comments.

Newcastle-Browne, shuffled on his chair. 'Bearing in mind Lord Richard's offer, this one from Brampton values the bricks and mortar at no more than £130,000,' he said profoundly. 'Worth more than that, I suspect.'

'Get on,' snarled Nickers.

The third envelope revealed an offer from Mr David Sedgefield of County Durham, who bid £305,000 for the whole farm. Enclosed was a long letter detailing how he planned to improve the holding. Payment, he claimed, presented no problems as the cash was already in the bank.

'Huh,' grunted Nickers dismissively. 'Next!'

'Next,' proclaimed Williams-Wetherby, cutting open another envelope, 'is an offer for the land south of the road, Lot 4, totalling 146 acres. It's from Mr Septimus Robson of Clartiehole. . . .'

'Aha,' said Nickers, 'old Sep eh? – Jolly good. He's one of our tenants, you know. Isn't he, Browne? Splendid chap . . . I bet he hasn't bid too much. . . .'

'£120,000,' announced the man from Spratt and Mackeral. 'That respresents about £822 an acre.'

'Interesting,' muttered Nickers thoughtfully. 'I'd rather like to see him get it . . . but that remains to be seen, what? Carry on,' he ordered.

The fifth offering came from Mr Douglas Donaldson. 'Oh yes,' said Nickers, 'the man with a scrum of sons, I remember. What does *he* say?'

'He's for Lot 1' said W.W. 'The whole thing. Says he intends to put his eldest boy in at Hindhope, and their price is . . . let me see . . . £295,000 . . . usual letter from his bank enclosed. . . .'

'Throw it in the fire,' said Nickers. 'That's no damned use . . . he's trying to steal the bloody place. Even old Filthy's offered far more than that! Move on!'

Mr Donaldson's bid was duly recorded however, and they waited for contestant number six.

'Interesting one,' said Williams-Wetherby, as he opened it up and scanned the contents. 'An offer for the farmhouse and the croft, Lot 3, from a Mr Gerald Wilberforce, Acacia Avenue, Newcastle. . . .'

'How much?' demanded Sir Nigel.

'£145,000.'

'Mm,' was all Nickers said. Newcastle-Browne cautiously raised an eyebrow, but contributed nothing else.

There were two other similar bids for Lot 3, apparently from city people eager to move out into the country. Both offers were made subject to a successful sale of their suburban homes.

'Huh, can't wait for those silly buggers,' snapped Sir Nigel. 'Next!'

'One more,' proclaimed the chairman; 'the last offer.' He took a little longer to cut it open. 'It's from that fellow Stevenson, builder chappie . . . he wants Lot 5, – house, steading and the land north of the road, about a hundred acres altogether.'

'His figure?' Nickers asked impatiently.

'£375,000,' said Williams-Wetherby.

'Good lord,' said Newcastle-Browne, looking down at his pad. 'That's more than old Filth – I mean Lord Richards – offered for the whole farm. . . .'

Les Stevenson's letter went slowly round the table, and came back to rest in front of Spratt and Mackeral's man. There were fully thirty seconds of silence. Eventually Nickers said, 'So that's it, what. Yes, very interesting. . . .'

'Well,' said Newcastle-Browne, 'in my opinion. . . .'

Nickers ignored him. 'I need a drink,' he said simply, 'and in *my* opinion we should remove to the County for a bite of lunch, and possibly an above-average bottle of something. We'll discuss these things later.' With that he stood up to indicate the meeting was adjourned.

All that morning Sep was on edge. It was not like him; he was usually very positive, dogmatic. Gladys would vouch for that . . . the original M.C.P. The man was *never* wrong . . . even when, to everyone else's satisfaction, he was ultimately proved to be mistaken. 'Ah well, the circumstances were extraordinary, were they not; some scoundrel had withheld a bit of vital information. . . .' He was likely to blame the weather, the government, the dog. Days later he might suddenly say, 'I still think I was right!' – and folks would wonder what he was on about.

But on 'deadline day' he flew about aimlessly, like a bird who has lost his tail feathers. He checked the sheep, but didn't really see them. He walked through the cattle, yet if one had been perched in a tree he might not have noticed. It was rather like driving from Newcastle to

Edinburgh without remembering any part of the journey. His mind was elsewhere. It was up at the Hall with his envelope.

Had he done the right thing? Had he offered too much? What if he hadn't offered plenty? How would he feel if somebody had beaten him by a thousand pounds? Relieved? Angry? T' hell with it, he would know soon enough.

However this didn't prevent him from paddling in and out of the kitchen every half-hour. He made a cup of coffee, and left it untouched on the bench. He went into the office, sat at the desk, opened and closed a few drawers, moved papers. 'Have there been any phone calls?' he asked, trying to sound disinterested.

'No,' said Gladys, 'nobody rang.'

'Has the postman been?' he muttered.

'Just a catalogue for me, and one of your farming comics,' she replied, moving his mucky wellies into the porch.

Once he came in and found Gladys chatting on the phone to Hilda, and his contrived tranquillity promptly disintegrated. 'Get off!' he screamed. 'I'm expectin' a call any minute . . . our whole future's in the balance, and you're discussing bloody cushion covers with some stupid woman! Hang up!'

Gladys quietly apologised to her friend, and promised to ring her back later. Then she moved his wellies again . . . outside the back door. It was raining.

Sep usually had 'forty winks' after his midday meal. His father had always done the same. 'After dinner doze a while; after supper walk a mile,' he'd said. Sep didn't always perform the second half of the rule (at least not during the winter) but the first bit was sacrosanct. Normally he would move from the table to the armchair, and begin to read the paper. Inevitably, after a few minutes, the paper would begin to fall lower and lower, along with his chin and his eyelids, until the whole body dissolved into a dormant, uncoordinated heap. He would groan, he would twitch, he might even mutter something unintelligible. If Sweep was lying nearby he would prick his lugs, raise an eye . . . but no more than that. Gladys would wash the dishes. After twenty minutes, as if a pre-set alarm had sounded in his head, the man would sniff, cough, splutter, and erupt into life again. 'Aye well, we'd better get on,' he'd say. 'Can't sit around here all day. . . .'

But on this day he didn't even clean his plate. 'What's the matter?' asked Gladys. He always ate everything, the last remnants usually mopped up with a slice of bread . . . really sophisticated.

'Nothin',' he said, 'nothin' at all. . . .' She knew well enough what the problem was, of course . . . she could read him like a book. An old, well-worn book.

Sep tried to doze, but his mind wouldn't lie down. He tossed and twisted in the chair. His left leg went to sleep, but the rest of him didn't. He read the mart reports (again) . . . lambs up a penny a kilo. Some gimmers and twins sold for seventy-five apiece . . . didn't sound very dear, but y' never know, they might have been rubbish. He would need some more sheep if he got the land on Hindhope.

He lit a cigarette and stared at the fire. He wished the waiting was over. He wasn't very good at waiting. His tummy rumbled, and Gladys at the sink smiled to herself. 'This isn't doin' m' indigestion any good,' he grumbled, '. . . I'm goin' for a walk.'

'Good idea,' she said, 'but put your mart wellies on, they're under the stairs . . . I think y' left the other ones outside. . . .'

After tea Sep and Willie fed the twenty bullocks that were still left in the hemmel. It was a routine they could perform with their eyes shut, saying nothing. A bag of meal poured along the trough, some hay thrown into the heck, baler twine hung on a nail, check water, doors and gates. If you listened you could hear the beasts munching. They would all be graded in a fortnight.

Later the two men were watching telly, waiting for the forecast, when the phone rang. Gladys answered it. On the screen John Humphries was reporting another explosion in Northern Ireland, another famine in Africa; a civil war here, a flood there, a revolution somewhere else. Nothing unusual, – just a level voice over coloured pictures from our man in Beirut, Belfast or wherever. It made you realise how lucky you were to live *here!* Nearer home a gang of youths had been arrested in a stolen car following a robbery in Sunderland.

'I know what I'd do with them,' muttered Sep, '. . . make them do a lambin' w' Cheviot yowes . . . that would slow them down!'

'It's for you,' said Gladys. She was standing at the sitting room door holding out the receiver. Sep cleared his throat, and came to take the phone.

'Who is it?' he asked.

'Mr Newcastle-Browne,' she said. 'He doesn't sound very well. . . .'

15

'The three wise men' had returned from their largely liquid lunch in fine fettle, and it was only after another hour of booming blather and loud laughing that they eventually got down to business again. Even then, Sir Nigel, his face aglow, his alcoholic breath liable to burst into flame at any moment, stomped around the sitting room like a loose horse. He caught a quick glimpse of himself in the gilded mirror above the fireplace, and saw fleeting images of Hanson, Goldsmith and Arthur Daley, – the other great wheeler-dealers of the decade. Now, with one financial master-stroke, he had despatched the covens of creditors baying at the gate. He could still be chairman of the Conservative Association. He had fashioned a reprieve for the Hunt. The Empire was saved!

'Bloody good show, what!' he roared, beaming down on his two companions. Williams-Wetherby was jubilant too. He looked up and saw a very satisfied client standing legs apart, hands on hips, his back to the fire. He lit a large cigar and, through the haze of smoke and gin, perceived a brightly coloured fruit machine registering his substantial commission. Fantastic! He even imagined his masters at Spratt and Mackeral offering him a seat on the board.

Newcastle-Browne was trying very hard to look over-joyed as well, but it wasn't easy with a blinding hangover. He should have known, – drinking with the dashing Sir Nigel in the middle of the day was a disaster scenario. Nickers was shouting now, and Newcastle-Browne's brain vibrated with every word.

Fortunately it did not take long to make the decisions. Nickers was quite clear what he wanted, and what he wanted was as much loot as possible. He didn't care where it came from, – class, style, were irrelevant now. Les Stevenson could have been a devout Liverpudlian militant and President of the Anti Blood Sports League . . . it didn't matter, – he would get the buildings and the land north of the road. What he did with it after that was his business. There was talk of development, even a golf course, – perhaps he wanted the farmhouse for himself, a country pad. Maybe he intended to be a farmer. . . . Who cared?

As for the rest of the farm, – well, it could not have worked out better. Here was a neighbouring farmer, a man of sound peasant stock, bidding a very fair price. What was more, he was already one of Sir Nigel's tenants, almost part of the 'firm'. Add his bid to Mr Stevenson's, and the total came to nearly half a million. Problems solved, roof repaired, the Canaletto stays on the wall. Daphne can have a new frock, the bank can go to hell!

'That's settled then,' shouted Nickers merrily. 'Time for a wee celebration drink, I think . . . What!'

*　　　*　　　*

Les Stevenson heard the news when Williams-Wetherby telephoned later that night. The conversation was brief and, after arranging to meet in the Spratt and Mackeral office next day, Les spent the evening getting his plans into gear. He began to dial numbers.

Charlie Woodruffe's wife was obliged to put her husband's supper back in the oven, as the planning applications were discussed. Clarence Shipley, the architect, was tracked down to an expensive restaurant, where he and Mrs Shipley were celebrating their twentieth wedding anniversary. Clarence was delighted to have the phone brought to the table by the head waiter, because (a) it made him look super-important, and (b) he had run out of things to say to his missus anyway, – and they were only on the 'starters'.

Les's solicitor had an answerphone. 'Sorry there's no one here to take your call,' the pompous voice declared, 'but if you leave a message after the squeaky noise, I'll get back to you very soon. . . .'

'Bugger that,' yelled Les. 'I know you're there, you legal creep, – pick up the bloody phone!'

The legal creep rang back within two minutes. 'I was in the bath,' he claimed. 'Been in court all day. . . .'

Les cut him short. 'We got it,' he said bluntly, – 'Spratt and Mackeral's office, nine tomorrow morning, see you there!' and put the phone down.

He talked with Thompson from the bank, with Hughie his foreman, and sat up late into the night considering how many men he could pull off other jobs to get the scheme started. It was tomorrow before he went to bed.

*　　*　　*

In Acacia Avenue there was some distress, – plaintive cries were heard coming from the Wilberforce residence. Prudence had not expected to hear the result so soon, indeed she hadn't really been thinking about it at all when the phone rang. She supposed it would be Delia overflowing with gossip and suburban scandal . . . a nice little chat before Gerald came home. Her heart missed several beats when Williams-Wetherby announced himself.

It was not absolutely necessary to inform all the runners so soon after the race, of course, but he was still flying on Nickers's high-octane hospitality, and he wanted to be first

125

with the news (good or bad) . . . image demanded no less. 'Frightfully sorry,' he said, and Prudence drooped immediately. 'Very good offer, almost made it . . . do keep in touch . . . bound to be something else soon. . . .' Prudence wasn't really listening now. 'Quite a few country houses coming onto the market these days,' he said. She heard herself say 'thank you,' and put the phone down while the man was still twittering.

Gerald found her sitting on the kitchen units by the sink. She was on her fourth gin, red-eyed, red-nosed, hiccuping.

'It's not the end of the world, darling,' he said.

'Yes, it is,' she sobbed. 'I'd set my heart on getting out of here (hic) . . . and the children had too. I hate this house, I hate the city. . . . We can't breathe properly, it's all your fault (hic) for being so frightfully mean. I bet another ten thousand would have done the trick (hic).'

'You're behaving like a spoilt little bitch, darling,' he protested. 'For God's sake pull yourself together . . . we can try again . . . I'll have a word with Spratt and whatsit later in the week. . . .' He put his arm round her wobbly shoulders. 'Come along,' he said, 'powder your nose and we'll go out to dinner, – drown our sorrows. . . .'

*   *   *

Sep's dialogue with Newcastle-Browne didn't last long either. Gladys heard him say, 'Uh huh, yes . . . right, aye . . . well right y' are then, uh huh, . . . g'night. . . .' That was it. The man's conversational powers on the phone were breathtaking!

'What was all that about?' she asked.

'It was Newcastle-Browne,' he said, going back to his chair.

'I know who it was,' she said, exasperated. 'What did he say?'

Her eyes were wide open, arms outstretched, appealing for information. Willie just waited quietly.

'He sounded bloody rough t' me,' said Sep; 'claimed he

had flu . . . probably been on the pop all day. . . .'

Gladys came to stand in front of him. 'What – did – he – say?' she asked very slowly.

'Oh, we got it,' mumbled Sep.

'God!' she screamed. 'You'd think it was third prize in a W.I. raffle, – a box o' biscuits, a pound of sausage. It's 140 acres, isn't it?'

'A hundred and forty-*two*,' he said. 'And a few trees. . . .'

'You're an awkward auld divil,' she smiled. 'Are we going to celebrate, then? How about dinner at the County?' she asked hopefully.

'We're not goin' there,' he snarled. 'It's full of nosey buggers wi' credit cards. . . . We might nip along to the 'Wheatheads' for a decent steak though, if y' like. . . .'

The four of them went: Sep and Gladys, Willie and Sandra . . . before the bar became crowded, a nice table in the corner. The ladies had fish and the 'House white'. . . . the lads demolished fillet steaks and a carafe of red. It was good.

'Make the most of it,' said Sep. 'We might not be out on the razzle for a while. We're in the red again, bonny lass.'

'Aye well, we've been *there* before,' Gladys said, waving for the waitress to bring more wine. 'Y' canna take it with y'.' She was becoming tiddly. 'Y' seldom see a hearse with a roof rack . . .' she giggled. The other three looked at her sideways.

Plates cleared, coffee served, Sep and Willie began discussing operation Hindhope . . . what to do, how to do it. They had been through the plans a hundred times before, but it was more than a hypothetical exercise now . . . no longer just scribbles on the back of an envelope or a fag packet.

The farming part was relatively easy. Here was this block of land divided into eight fields, each of which might require different treatment, different management. Some of it was sown with cereals already. There would be livestock to buy for the grass, extra fertiliser and sprays needed. But that aspect of the 'adventure' didn't worry them too much

. . . commonsense and experience would ensure they did most things right. It was the borrowing that bothered Sep.

At that moment, sitting there in the restaurant, relaxed, nicely full, a brandy in one hand, a cigarette in the other, chatting happily with his family, thinking how attractive Sandra looked in her little black dress . . . at that moment, he still had money in the bank. He was comfy. It hadn't always been like that.

He didn't need a Mastermind memory to recall earlier days, touch-and-go days, sometimes. When the kids were little and socks had to be darned. Selling a load of barley, a couple of bullocks, a dozen lambs, – not necessarily when he wanted to, but because he *had* to. When the overdraft temperature pushed up the bubble on the bank thermometer. Happy times, though. They had worked their way through that territory (with a little good fortune on the way, – y' needed it). Those were days when the farming ladder still had rungs at the bottom . . . they seemed to be missing now.

'Still time t' change your mind. . . .' Willie maybe guessed what his father was thinking about.

'No, it's the right thing t' do,' said Sep, stretching and rubbing his tummy. 'If y' stand still you'll be run over.' He made a sign for the lass to bring the bill . . . his left hand a notepad, his right hand a pen pretending to scribble 'the damage'. 'We'll sign on the dotted line tomorrow,' he said. . . . 'I'm ready for bed.'

There were still a few lights on in the village as they drove home, and they passed the Forsythe lads walking back from the White Hart. Sep couldn't hear him of course, but Samuel was telling his brother (for the umpteenth time) how relieved he was to be out of farming. 'Best thing we ever did,' he said, 'another year would've killed us.'

'You're probably right,' Thomas mumbled, 'but I think I might miss it a bit come next spring. . . .' A few paces on he added, 'In fact I might take a lambin' . . . just t' keep m' eye in, y' understand.'

'Y' must be crackers,' said Samuel.

16

The country quilt was dark green and startling yellow.
Grass fields full of lambs and freed 'hemmel hostages'.
Silage stuffed into big black bin-liners. Sprayers drizzling
along barley tramlines. The Australian tourists were in the
field at Trent Bridge, and a mild frost touched the leaves
on Fred Little's geraniums . . . as if to remind everybody
who was in charge.

At Hindhope, Les Stevenson's army was on the move.
On Day One they knocked down Malcolm's old tin
barn, and began to demolish the new shed beside the
farmhouse. Two dumper trucks, a JCB and ten strong
young men with tattooed arms and wheelbarrows had
begun to gut the farm buildings. Already there was an
artist's impression of the finished development: *Spacious
country homes for the discerning executive family*. En-
suite bathrooms with bidets, fully fitted kitchens with
dishwashers, and three separate pony paddocks for little
'Desert Orchids'.

The lads removed the roofs, carefully piling the slates
and tiles for re-use on the new buildings. They cleared

the debris, burned the rubbish, unloaded bricks and breeze blocks, sand and cement, – preparing the way for brickies, plasterers, chippies and electricians.

Clarence Shipley scurried about with his drawings and a tape measure. Charlie Woodruffe looked in to ensure Arthur was not secretly developing another Disney World. Building inspectors arrived unannounced to check materials and workmanship. Thompson came to see where all the money was going.

At the White Hart a considered debate on the venture came down heavily in favour of 'a cock-up'.

'They're talking Monopoly money,' declared Geordie Dodds, £200,000 for each house . . . three houses, that's er. . . .'

Willie Turnbull helped him out, '. . . six hundred thousand,' he said.

'Well there y' are then,' said Geordie, looking admiringly at the mathematician. 'Where's that sort o' money comin' from . . . eh?'

'Three solicitors,' suggested someone from the domino board, 'and I'm knockin'. . . .'

By mid-summer, Les Stevenson was in danger of knocking as well. He had a lot of money lying out now, and some of it was costing 19 per cent. The price of materials and labour seemed to rise every week. He had taken men away from other, smaller jobs, and upset other clients. A house sale he'd been banking on had not materialised. He had lowered the price twice . . . still no takers. There was simply no money coming in now . . . and the lads were all on overtime. The VAT man, the tax man, the Safety Inspector were all knocking at his portakabin door. Thompson from the bank (who only a few months ago would have willingly, eagerly loaned him enough cash to take over Wimpey) was now muttering about him being 'over-extended'. He thought property prices had probably peaked. He said Head Office were a little less 'bullish'. It was a warning.

Les decided he would have to sell something pretty

soon, or there might be a danger he would run out of finance before he completed the job. He was suddenly sailing very close to some nasty rocks.

Relief came in green wellies and yellow dungarees. It was a bad Monday morning. A very wet night had transformed the site into a sea of mud, – drainage ditches and septic tank cavities were full of water. The last thing he needed was a twittering woman following him about like a stray dog . . . getting in his way. She was probably from the Royal society for the protection of homeless hedgehogs, or tormented toads, or harassed hares.

He looked back irritably as the lady stumbled over a pile of stones, almost losing a wellie. She was waving at him. 'A wildlife lady if ever I saw one,' thought Les, 'I'll get a lecture on the rape of the environment. Or maybe she's a parish councillor concerned about access roads and sewage disposal. . . .' She looked like nothing but trouble. Well, he was used to handling that, – he would get rid of her with the first salvo.

'What the hell do *you* want?' he roared, 'I haven't time to chat to any ramblers who've strayed from the footpath, – and I'm not buying anything . . . and I'm not filling in any questionnaires . . . so bugger off, Madam!'

'Morning,' she smiled sweetly, completely unperturbed. Perhaps she hadn't heard, '. . . Mr Stevenson?'

'Look,' said Les, 'just go away before you're bricked up or buried. We're busy!'

'Of course you are,' she said, still smiling. 'But I wondered if you might sell me some land. . . .'

'What?'

'Yes, well I'm not terribly sure how much . . . perhaps ten acres. . . .' She lobbed the figure at him, and he caught it.

'Ten acres, eh?' He was never at his best on messy Monday mornings.

'Well, to be perfectly honest,' she said, 'I'm not absolutely sure what an acre is . . . can't actually visualise it. We'd just like somewhere to keep a few animals . . .

some ewes perhaps.' (She pronounced ewes as yooze.)

'Come into my parl . . . er office,' said Les. 'I think we might be able to help. . . .'

Half an hour later, a delighted Polly Pillick had purchased all of the East Haugh, eighteen acres of old pasture at twelve hundred pounds an acre. 'Always wanted to be a farmer,' she grinned as she left the Portakabin . . . 'soopah!'

'Absolutely,' said Les, and he took the cheque straight to the bank.

Perhaps it was this little deal that concentrated his mind; perhaps it was the condition of the site that kept him in the office, staring out of the window . . . thinking. Whatever it was, he phoned Williams-Wetherby before he went home.

'The farmhouse?' said Williams-Wetherby. 'You'd like to sell the farmhouse? Well of course, dear chap . . . I'm sure we can assist . . . in fact I think I know someone who might be very interested. They may want some land as well, however . . . they have a few rural delusions, you understand. . . . A little plot in the country, dogs, horses, a few sheep, a hen or two . . . you know the sort?'

'Oh indeed,' said Les simply. 'To be honest, I even considered renovating that house for myself, maybe retiring there when all the work was finished to breed a few Highland cattle or some such thing. Anyway,' he went on, 'I thought we might test the market . . . see what it's worth. . . .'

'And the land?' enquired Williams-Wetherby, already savouring the prospect of selling it twice. 'I'm sure the people I have in mind will want some land.'

'No problem,' said Les. 'There's about seventy acres left. I'll consider any reasonable offer . . . no hurry y' understand. . . .'

'Oh, I understand,' said the voice at the other end of the phone. He had already found the number of Gerald Wilberforce. 'We'll be in touch.'

132

Williams-Wetherby pondered whether to ring Gerald at his office, or Prudence at home. She would probably be in the garden. He imagined her in a headscarf, a chunky sweater, green overalls and gloves, – trowel in hand, planting high-priced blooms from the garden centre. She might be kneeling on an old plastic floor tile . . . it would be green as well.

'Double-five, seven, double-one,' she said breathlessly, removing the other gardening glove.

'Ah, Mrs Wilberforce,' gushed Williams-Wetherby in his best 'des-res' voice, '. . . we have some news you may find interesting, – it concerns Hindhope. . . .'

Two days later Gerald and his wife drove to the farm. Approaching from the south up Rabbit Lonnen they came upon what was clearly an agitated country person kicking a sheep as it lay on the grass verge. It seemed the sheep was totally indifferent to the assault. The man was carrying a lamb, perhaps belonging to the victim, and several other ewes and lambs were blocking the road. A collie dog was hiding behind a tree.

'Excuse me,' said Gerald, as the electric window glided down. 'Do you live around here?'

Sep considered telling the idiot in the BMW that no, he was in fact a nomadic Kurdish sheep stealer on his way back to the mountains, – but he caught a glimpse of a lady in the passenger seat, leaning over and smiling sweetly. He thought he'd seen her before somewhere.

'Yes,' said Sep, 'we live just down the road, farm this land here. . . .' He gestured vaguely over the hedge. 'Are y' lost?'

'Oh, we're certainly not lost, are we, darling?' Gerald smiled at his wife. 'Actually we may be buying Hindhope . . . or what's left of it. . . .'

'Really,' said Sep thoughtfully. The yow had miraculously recovered and rejoined her companions, – bleating for her offspring. The improved atmosphere encouraged Sweep to emerge from cover and put his paws up onto the gleaming car door. He poked his silly, grinning face

133

through the open window to be patted by Prudence, and drooled on her lap.

Gerald got out to introduce himself and shake hands. He noticed it was a very large hand. Sep put the lamb down, and it limped off in search of mother. Prudence was out of the car 'oohing and ahing' . . . her thighs already covered in clarts by Sweep's filthy feet, – but she didn't seem to mind.

'I wonder,' she said, 'if we could ask this gentleman's advice, – after all, darling, he *is* a proper farmer, and we might even be neighbours.'

'He may be in direct competition, of course,' Gerald pointed out.

'Not me,' said Sep. 'I've got plenty on m' plate. . . .'

They stood there on the road as the ewes and lambs grazed, Sweep sometimes turning a 'wanderer' back to the bunch, and talked about the farm, the land and Les Stevenson's development. After about twenty minutes Sep said he would have to go, '. . . nice to meet you,' he said. He wished them luck; they seemed nice people. They hadn't a clue, 'but nea harm in them', he thought.

'What was the matter with that sheep?' asked Prudence, 'the one you were . . . er . . . the one that was lying down. . . .'

'Oh that's a very common sheep disease,' said Sep wisely. 'You'll likely become familiar with it when you have some of your own. . . .' He paused. 'It's a mental problem,' he said, 'it's called bloody awkwardness. . . .'

Up at the farm Les was waiting for them, chatting to Williams-Wetherby who introduced everybody, and then retreated to his Land Rover. 'Let them get on with it,' he thought. 'Whatever the result, I win!'

Gerald got down to business straight away, and that suited Les. Prudence disappeared into the farmhouse. She'd been given instructions to keep out of the way. In the Portakabin, Les lit a cigar and declared he'd accept £235,000 for the lot: the old house and about seventy acres.

134

'Ridiculous,' said Gerald.

'C'mon,' snorted Les. 'The house is worth a hundred and fifty grand of anybody's money!'

Gerald sighed. 'Possibly,' he said, 'but only after someone's spent £50,000 making it habitable. There's no damp course, all the window frames are rotten . . . and, well, I could go on, but you'll be perfectly aware of these things. We had it surveyed and valued by a very respected firm when we first offered for it, and their price was certainly considerably lower than that.

Les was taken by surprise. That bloke Wetherby-Williams (or whatever his name was) said the fella would be a push-over . . . rich city fairy he said, . . . didn't know his arse from his elbow. . . . Les's voice was a little too high to be entirely convincing when he said, '. . . the land's worth a thousand an acre. . . .'

'No it's not,' Gerald corrected him. 'Six hundred would be nearer the mark. Camp Hill has no depth of soil at all, there are stones everywhere; none of it can be cultivated satisfactorily, other than at enormous expense . . . limited usage. Much of the land is poorly drained . . . major scheme required to bring it into full production. . . .'

Les was trying to interrupt, but it wasn't easy. 'Furthermore,' said Gerald, 'the pH is down at 5.2 and the fencing is nowhere near stock-proof, particularly along the Hind burn. I could go on to point out that, historically, livestock have seldom done particularly well on this land, possibly due to a distinct lack of lime and phosphate, and a magnesium deficiency.' (He barely paused to draw breath.) 'As well as a substantial expenditure required on drainage, fertiliser and fencing, I am advised that the price of the livestock required to graze this land will now be at a seasonal high . . . and consequently any return on capital will be derisory, probably lower than the Gateshead Building Society Instant Saver Account. However,' he was on his feet making for the door, '. . . my wife and I, after careful consideration, and learned counsel, some of it taken

135

within the last half-hour, have decided to offer you £200,000!'

'Where did you get all that stuff from?' stuttered Les. His cigar was out. This had not gone according to his script.

'One more point,' said Gerald, holding the door open. 'We would be happy to complete within the month, provided we have access immediately. No doubt you'll make a decision quite soon. . . .'

'What?' was all Les managed to say as the door closed.

Williams-Wetherby found him poring over bank statements, a calculator responding to fingertrip instructions, another cigar smouldering in the ashtray. The telephone was ringing, – Williams-Wetherby knew when to make a strategic withdrawal.

Les did his sums again, talked to Thompson again, to the solicitor and the accountant again . . . and finally phoned Gerald at his office on Friday, accepting his offer. Down at Spratt and Mackeral, W.W. was jubilant: Hindhope, a very average farm of 245 acres, with dilapidated farmhouse and antiquated buildings, had been sold for about double the going agricultural value. What was more, within a couple of months he had sold almost half of it all over again!

'Not bad,' he smiled, 'not bad at all.' He opened the bottom right-hand drawer of his desk and took out a bottle of Chivas Regal. Well, everyone was happy, weren't they? That fellow Sep was happy; he now had a farm big enough to be viable, he said, something for his son to build on. Gerald what'sisname was happy, and his wife was on cloud nine. Mrs Pillick was over the moon with her field.

Les Stevenson was happy: he had managed a nice little asset-stripping job . . . kept the cash flowing, financed his development. He should end up with a good profit. And Thompson at the bank was happy; the fellow had been a little worried for a while, but it was all coming right now. He would be preening himself. And we

136

mustn't forget Sir Nigel, – he was overjoyed at the deal. . . .

Williams-Wetherby cupped the glass in his hands and watched the golden nectar swirling gently. . . . 'Yes,' he smiled, 'I do believe everyone should be delighted!'

17

The changes at Hindhope created waves throughout the neighbourhood. The population up at the farm rose dramatically. It became a commuter's commune, complete with big cars and small hairy ponies, fed on Smarties. A swarm of self-confident children in hard hats called Jack, Victoria, Caroline and Rupert 'evented' in the croft at weekends, while daddies laid flagstones on their patios.

The old steading was now 'Hindhope Court', three luxury homes around a cobbled courtyard. The Gaskett family were in No.1 (Mr Gaskett a second-hand car dealer); the Barrister was at No.2, and Beeday the plumber in No.3. (Barry Beeday preferred to call himself 'a sanitary consultant'.)

Tommy Cleghorn was in demand to 'landscape' the virgin gardens. Quickly christened 'Capability Cleghorn' by the newcomers, he planted leylandii, laid lawns, and did his best to resist the temptations of the barrister's wife, who served him coffee and chocolate wholemeal biscuits while still in her nightgown.

The daily traffic through the village doubled as the city

folk drove to and from work, school or supermarket. One morning, when there was a snarl-up at the war memorial involving the school bus and the postman, Geordie Dodds reported a six-car tail-back as far as the Forge. 'It's worse than that M25!' he said.

Peter Foggin at the garage was not complaining. Four new petrol accounts, with customers who actually paid at the end of the month. 'This could be the beginning of boom-time,' he said. 'We have lift-off!'

At the White Hart Jack and Nora briefly considered building an extension out beyond the gents to serve Scampi Provençal and Black Forest gâteau in 'sophisticated surroundings'. When Gerald and Prudence strolled into the bar one night, ordering 'very dry Martinis, and a little liver paté perhaps', mine hosts even wondered if they should invest in some fine wines as well. In the end they settled for a bigger toasted sandwich machine, a much wider selection of crisps, and half a dozen bottles of Muscadet from Safeway's. But they knew life would never be the same again.

The Pratts, Peabodys and the Pillicks all considered themselves members of the agricultural community now . . . albeit still on probation. Nevertheless, they visited the mart at least once a month (dressed as if for the opera) and talked knowledgeably about 'staggers'. They regularly gathered at each other's homes for exclusive supper parties to discuss the EEC, their salaries in the city, the proposed withdrawal of the lamb premium, their index-linked pension schemes, the price of fillet steak, holidays in Tuscany, the excessive use of pesticides, the brilliance of their children . . . and convinced each other how environmentally aware they all were. They read *Country Living* and *Big Farm Weekly*, – sometimes even rose early to hear *Farming Today* on Radio 4.

As an accountant, Trevor Pratt at the Forge knew only too well that Naomi's farming fantasies on 1.65 acres did not make a lot of sense. But what could you do with someone who persisted in dressing up like an Afghan tribeswoman to feed two ponies, seven pedigree Jacob ewes, – and now a garage full of sex-crazed rabbits. When the Jacobs eventually gave birth to ten nondescript lambs, with few of the breed characteristics one might expect, Geordie from next door was consulted. 'Aye well,' he said philosophically, 'looks like y' might've had a lodger in. . . .'

The Peabodys at Paddock House had bought three heifer stirks, and now referred to their 2.4 acre field as 'the Ranch'. This move into cattle (they felt) placed them on a much higher plain than the pitiful Pratts with their miserable sheep. Almost everyone they knew had at least one sheep. Sheep were common and rather smelly. Cattle on the other hand, especially when viewed from the sitting room window while sipping a sherry, looked . . . well . . . bigger.

Geordie reckoned the Peabody's heifers would *never* be any bigger than they were now. 'Short-coupled, badly bred little buggers,' was his assessment. When Mrs P began twittering about the nucleus of a herd, and maybe buying a limousin bull, he was forced to retreat into the hayshed as the tears rolled down his cheeks.

Giles and Polly Pillick referred to their newly acquired East Haugh as their 'small estate'. 'We have a small estate in the country,' Polly would tell her city friends, smiling, eyes closed, waiting for gasps of admiration. 'And we're totally organic, of course; no narsty fertilisers or chemicals. It's all quite lovely . . . the meadow, the river, the birds. . . .' One almost expected her to warble a few bars from *The Sound of Music*.

'That's all very well,' somebody would ask, 'but what on earth do you do with all that land . . . presumably you have a manager?'

'Actually we've discoverd *the* most wonderful fellow. . . .' Polly sounded as if she had recently come across Geordie on a shelf in Peter Dominic's; '. . . a vintage character,' she would say. 'So colourful, mellow, immensely patient, and quite brilliant with animals!'

Indeed Geordie had taken on the mantle of a rural guru. This sixty-two-year-old, seasoned smallholder with a wonky knee (twisted at the dippin' 1972), half the index finger missing on his left hand (pullin' turnips 1958), mangled big toe (bullock stood on it in '61), and a dodgy back (cowpin' a muckle tup at the clippin' three years ago) . . . was the man these new instant farmers turned to in times of stress. In any other business he would have been called a 'consultant' and charged a hundred quid an hour. . . .

'Cilla is unwell.' Mrs Peabody was almost suicidal with worry: one of her heifers was coughing and wheezing.

'Worms,' Geordie diagnosed, and on the Sunday morning he would walk the three rubbishy stirks across the road, push them into his cattle crush, and dose them without much fuss.

A distraught Naomi Pratt came running into the yard one day cuddling a very wretched lamb, with cold, droopy lugs and dry, slack skin. 'It's not well,' she sobbed. 'I think it's going to die. . . .'

'Very probably,' said Geordie. 'As a rule, if y' get nowt t' eat, – y' die. This poor little sod's starved!' Further investigations revealed the patient's twin to be as full as an egg, and a mother with mastitis.

'She's only firin' on one tit,' declared Geordie. 'Did y' not see the bitch draggin' a leg?' He gave the penicillin injection, guessing it was probably too late, and withdrew, mumbling about 'useless pin-stripe peasants.'

The Pillick outfit was on a different level altogether. Giles and Polly swung into action with the passionate enthusiasm of enlightened evangelists. Here in the rural jungle, where for years the natives had been persuaded to worship a God who promised four tonnes an acre and a 200 per cent lambing crop (provided they made sacrificial offerings of nitrogen, pesticides and sundry medicines) . . . here would be an oasis where mother nature could rediscover herself. Here the Pillicks could wander about up to their intellectual armpits in a meadow made for missionaries.

'We'll put it all into set-aside,' announced Giles, 'and get enormous amounts of money from the government. All the old weeds will re-emerge; it'll be full of bluebells and badgers . . . wonderful. . . .'

'It doesn't qualify for set-aside,' said the man from the ministry.

'How about an Environmentally Sensitive Area?' asked Polly. 'I believe we could be paid hundreds of pounds per hectare if we dammed the river and turned everything into a water meadow.'

'No chance,' said the man from the ministry, thinking to himself how mercenary some missionaries could be. When they asked for details of the Countryside Steward-ship Scheme he put the phone down and went for lunch with Alison from 'accounts'. He had long admired Alison's areas of special scientific interest.

It took the Pillicks a month to decide what to do. Luckily they were rich enough to do nothing for quite a while. As chairman and managing director of Applepeel Electronics, Giles had paid himself several substantial bonuses over the past few years, profited greatly from some timely share transactions, and negotiated a very generous pay award and pension scheme. He was a very successful businessman. He had an MBE from Mrs Thatcher to prove it.

It was David Attenborough who finally guided them out of the 'wildnerness'. There he was on the telly one evening, crawling about in the steaming rainforest, ear-nestly whispering something about the mating habits of the two-toed tarantula. '. . . only seven of them left in the whole world,' he breathed, 'habitat destroyed, numbers decimated, almost extinct. . . .'

'That's it,' exclaimed Polly, 'that's what we'll do with our estate. . . .'

'Spiders?' Giles was unconvinced.

'No, not exactly,' said Polly. 'Endangered things . . . a

kind of sanctuary for all sorts of unfashionable animals . . . rare breeds . . . there'll probably be an EEC grant available. . . .'

That's how Geordie Dodds found himself delivering feed to two guanaco llamas, a pregnant Vietnamese pot-bellied sow, half a dozen Hebridean sheep (Geordie thought they were goats until he saw the real thing . . . a pure-bred Angora nanny complete with kid) and a dozen little black Dexter heifers. 'Even smaller than those rubbish at the "Ranch",' he said.

And this was just the beginning. The Pillicks stopped taking *Farmers Weekly* and *Power Farming*, and subscribed instead to *Ark* and *Fancy Fowl*, and tuned into a *A Small Country Living* on the radio. They travelled all over the country to buy 'Millefleur' bantams, an Exmoor pony, a Bagot billy . . . a superannuated donkey . . . until Haugh House became known locally as 'The Menagerie'. Cars and buses stopped on the road as wide-eyed travellers took photographs. Walkers wandered off the footpaths to catch a glimpse of this exotica.

Such was the interest that Polly even considered providing afternoon teas and Range Rover safaris, perhaps opening a shop to sell rural crafts made by the local country people. However, when the pot-bellied pig fell asleep on her new-born litter; when the pony was seen shaking a Hebridean lamb in its jaws; when their Jack Russell terrier slaughtered several careless guinea fowl; when the llamas escaped and attacked a W.I. bus trip at the White Hart, Polly realised 'farming' could be full of surprises, and very often visitors were the last thing you needed!

Geordie wouldn't admit it, but he was just a wee bit impressed. A lifetime in the business, and he'd never seen nor heard tell of these obscure animals. Most of them, he reckoned, were of no commercial use and should be in a zoo . . . but it was no skin off his nose. He'd happily deliver bales of hay and straw to the prosperous Pillicks for as long as they paid him cash money.

Up at Hindhope the farming was more conventional. Gerald had been happy to leave Prudence in charge of house renovations and the master agricultural plan, while he did his best to earn enough in the city to pay for it all.

Luckily, his wife realised at an early stage she was way out of her depth. So far it had only been dreaming; now it was for real. Where does one begin?

She began by shrewdly charming the wellies off Sep. And Sep, who hadn't been so seriously chatted up since Smithfield Show, 1965 (when he was dragged into an underground strip club by a posse of wealthy Yorkshiremen), gently melted.

'I didn't really want to get involved,' he told Gladys, 'but the lass hasn't a clue. . . .'

'She's bonny,' said Gladys.

'Do y' think so?' Sep tried to sound as if he'd never even considered such a thing. 'Aye well, I suppose so,' he conceded, 'in a townie sort of way, if y' know what I mean. . . .'

'Wears tight jeans and little T shirts,' Gladys smiled.

'Never noticed,' said Sep.

'Liar,' said Gladys, and started up the vacuum.

Sep had taken over the Middlewhite field, Rimside and the Happorth, all in winter barley. (They'd have to do something about the wild oats in Middlewhite.) The front field was in wheat. That still left another sixty-four acres to stock this year, and as soon as possible, he left Willie to get on with everyday things and went to the marts. He was buying for Prudence as well.

On her side of the road there was a crop of barley in the West field (she could keep some for feed, and cash most of it) while Camp Hill, West Haugh and the Lambing field were full of grass. There'd been nothing on them all winter, and they should carry maybe three ewes with twins to the acre, plus a few cattle. That would be enough to keep the lady busy, – out of mischief. Sep imagined profit was not a major consideration; he certainly hoped not, because there wouldn't be any.

144

He bought 125 mule ewes for her, all with pairs, and fifty store cattle to finish by the back end. The sheep looked worth the money, but the cattle were no bigger when he got them home. 'Dear enough,' he thought. 'They generally are. . . .'

Prudence, however, was delighted. She even purchased a half-worn collie from Willie Turnbull, went on a weekend sheep management course at the Agricultural College, – and wandered round the fields at home talking to the animals and feeding chocolates to the devoted dog.

Gerald wasn't quite so thrilled at events. In fact he became increasingly dismayed at the Hindhope bills. The house in Acacia Avenue hadn't attracted a buyer yet, either . . . and prices were going *down*. Now he was saddled with a flock of sheep and a lot of cows (they were all cows to Gerald). How the hell did he get into this?

The executive residents in Hindhope Court watched with mixed feelings as the farm came to life. In the beginning most of them viewed the livestock with innocent curiosity . . . the children especially, enthralled by cuddly ewes and lambs. They were more wary of the cattle, perhaps some of those big, brown creatures were bulls. The big, brown creatures would come to view the viewers over the fence with a similar innocent curiosity . . . to sniff and lick outstretched hands. Prudence was invited in for coffee to explain the mysteries of farming life.

'You'll sharp get sick o' them,' warned Sep, and the first indication he might be right came when they dosed the lambs.

'What on earth are you doing?' asked one of the 'courtiers'.

'Worming them,' grunted Sep, as he thrust the dosing gun down a throat. 'Fetch your kids over and we'll give them a shot . . . I think that little 'n' might need it. . . .' The mother was not amused, and the 'little 'n' ' fled in terror.

They all came to watch the clipping, pointing and giggling as naked yowes ran off bleating for their lambs, leaving a pile of wool behind. They were impressed how quickly Willie and his mate Alfie could disrobe a sheep. Alfie was built to resemble a small byre. He moved like a combination of John Wayne and Terminator 2, handled yowes as if they were paper bags, and swore with a vocabulary Prudence was unfamiliar with. When a lean, fidgety beast with no rise presented a few problems, and eventually left with several bits missing, followed by some colourful language, – the audience began to mutter and shuffle. When Alfie removed half a lug from another awkward sheep there was blood everywhere; the plumber's wife from No.3 fainted and had to be helped home for a brandy.

They weren't thrilled when dipping time arrived either. A strong west wind prompted complaints of 'atmospheric pollution' . . . 'smelly poisonous substances threatening the lives of children and other wildlife. . . .' Somebody even mentioned Chernobyl. Prudence, peering out from new yellow oilskins, tried to assure them that the bath was medicinal, necessary, required by law . . . but the barrister's wife said her delphiniums were dying.

A few days later it was more than the lady's delphiniums that were in trouble, as a heifer came through the fence from West Haugh, plunged a recently turfed lawn and crapped all over the patio. The barrister talked of litigation. Prudence apologised profusely, promising to repair any damage.

Sep declared it was an act of God. 'The beast was mad a-bullin',' he said.

Prudence didn't want to spray the barley at all. Not only because it might endanger more blooms, but . . . well, she was not in favour of chemical sprays, anyway. She was convinced the operation would wipe out all forms of wildlife, creep into the water supply and sterilise the local population, and probably make the hole in the ozone layer even bigger.

Sep felt obliged to walk her through the crop, pointing out the mildew and Ryhnchosporium, and the inevitable effect on yield if left untreated. He brought in the local expert rep from Technochem who (naturally) painted an even grimmer picture, an all-consuming plague of biblical proportions unless controlled immediately.

Gerald needed little persuasion. 'For God's sake spray the stuff!' he said. 'If that field's a failure, we're finished. We need a record harvest as soon as possible! Couldn't we combine it tomorrow?' he asked desperately.

'Not really,' said Sep, '. . . it's still green.'

18

The likes of Sep, Geordie Dodds, Tommy Cleghorn's mother and a few other 'old blooms', could remember when the Village Hall was used most nights in the week. Right through the winter there would be highly competitive whist drives with maybe fifteen tables.

The N.F.U. always met there. Hill farmers would Land Rover down from the fells, still carrying their sticks, to complain about cow subsidies, the high price of barley, and commission at the marts. Now they mostly 'Shoguned' to the Cheviot Suite at the County, and complained about lamb prices, Less Favoured Areas and silly EEC rules and regulations. The big cereal growers used to swop yield exaggerations and moan about the merchants. Now they compared bushel weights, nitrogen levels, protein content . . . muttered angrily about co-responsibility levies, and all the paperwork they had to do.

The sheep men still wander in with that bewildered expression sheep men often have, talk wistfully of days when the Ministry sent a cheque after every mart, and compare improbable tales of death among their flocks.

'Y'll not believe this,' says Willie, 'but we had a lamb ate a mole trap last week. Found it dead as a maggot with the trap stuck in his throat . . . choked t' death . . . a right good lamb 'n' all!'

'That's nowt,' declares Charlie, unimpressed. 'Y' know the chain on a hay heck . . . it has a hook on the end t' fasten down the lid, – know what I mean? Well, we had a lamb swallowed the hook . . . like a bloody suicidal salmon. There it was stone dead yesterday mornin'!'

'Lost a canny lamb in the croft on Monday,' says Arthur, not to be outdone. 'The wife had just taken the washin' in, wrapped up the line and tied it t' the pole. This brainless lamb must've jumped up and got his head fast in the loops, twisted the rope round his neck and hanged himself. . . . Brilliant, eh?'

The W.I. still met in the Hall until quite recently, but they say it's too cold for them now. It cost a fortune to put the heating on, and even then, their thermals hardly coped with the north-east draughts.

Once upon a time Young Farmers danced there, – hurled their future wives about as if they were wayward gimmers.

'Take your partners for the Bradford Barn Dance,' . . . and the sexes would circle, eyeing each other cautiously, waiting for the music to stop . . . hoping desperately it stopped when you were opposite the bonny lass in blue. Please God, – not the fat, smelly one from Berwick!

'Everybody on the floor for the Dashing White Sergeant' . . . join hands, swing and swirl, yell and shout, sweat and show off. Shy, spotty youths and overweight wallflowers sit it out, smiling bravely through suicidal depressions.

Smoochy Sinatra love songs, twisting Rolling Stones, girls dancing together, handbags on the floor. Yesterday's steps and moods and fashions . . . and after the last (Tennessee) waltz, they roared away in old vans and pick-ups to steam up the windows, parked in some secluded gateway.

149

'Do y' think we should get married, pet?' he might ask casually.

'I wouldn't mind,' she replies, rearranging herself. 'When do y' fancy?'

'Ah well,' says Romeo, 'it canna be for a while yet 'cos we've still got thirty acres o' wheat t' sow. Then there's the tups t' put away . . . and the calf sales. . . .'

Ah yes, sex and the single farmer was always a passionate affair.

These days, Peter Foggin from the garage organises a folk night for the Hall funds. 'Just to keep the roof on,' he says, – and the Percy Elliot Three do their best to lift it off. Percy plays one of those electronic keyboards, his son is on the drums, and Elsie Patterson wrestles with her accordion. It's not easy for Elsie, she'd be the first to admit it; she's a big lass with a mountainous chest. Consequently she's always been obliged to play the instrument at arm's length (well, you can imagine the dangers) . . . but she's a good player.

Herbie Flood, who lives in the old schoolmaster's house and works for the Water Board, gives a turn on the Northumbrian pipes. It's generally 'Waters of Tyne' and 'Keep your Feet Still Geordie Hinney', with 'Cushy Butterfield' as an encore, if the audience shows any enthusiasm.

Jack and Nora from the pub provide a bar in the back room, so people are usually in a very benevolent mood by about nine o'clock.

An old bandy-legged shepherd from up the Coquet valley climbs onto the stage, takes a mouth-organ from his coat pocket, knocks the crumbs and fluff out on his knee, and plays 'The Muckin o' Geordie's Byre', and a selection from *The Student Prince*. He never had a lesson in his life. A plumber from Newcastle plays mad hornpipes and plaintive ballads on the fiddle . . . and Fred Little from the shop does a Bobby Thompson impersonation, complete with cap and bent tab.

Peter is the M.C. Between acts he tells the kind of

stories that make old ladies shriek and squeal, as if they'd never heard such things before.

The night generally ends with the delectable Gloria Swanson from West Cottage singing 'Abide with Me'. . . and most of the lads there wouldn't need to be asked twice.

It's always a sell-out, and the roof rendered weather-proof for another few months.

The Parish Council met at the Village Hall the evening before the 'Countryside in Bloom' judging. Hindburn had entered in the small village class, and after being placed third the previous year, the committee was determined to win this time. Everybody had done their bit. The verges were cut and trimmed, every garden was tidy and aglow; the grass in the graveyard mown like a lawn; bird droppings scrubbed off the war memorial. Everywhere there were stone troughs and hanging baskets filled with pansies, primulas and petunias, antirrhinums, sweet William and lobelia. Even the road had been brushed from Paddock House to Haugh House: 'you could eat your dinner off it.' Mrs Peabody had cleaned the old red telephone kiosk till it shone like a beacon.

Mrs Graham at the vicarage would meet the judges at ten on the big day, serve them coffee, and conduct them through the delights of the village. If it was a sunny day, surely they'd be impressed.

It must've been about five in the morning, broad daylight, when fourteen Hebridean ewes with lambs, six Dexter cows, a goat and a pot-bellied pig sauntered out through the Pillicks' garden and onto the road. If they'd then turned right and wandered westwards it might not have mattered too much; however, they went left towards the spotless village.

The pig waddled no further than Burn Cottage where Thomas Forsythe, standing at the kitchen window, was surprised and disappointed to see the beast devouring cabbages in his back garden.

By this time the Dexters had grazed their way slowly

down the road as far as Geordie Dodds's farm. They would've gone further, but Geordie, having an early morning smoke on the doorstep, saw five cattle go past his gate. He scrambled over the garden wall and confronted them at the war memorial. Luckily Tommy Cleghorn came out of Glebe Cottage at the same time, and the two men managed to herd the cows off the road and into Geordie's croft. They were unaware that the cattle had already been up the lane by the side of the school and that one of them was still there. Now she found herself alone, and after briefly considering her position, she took fright and came blaring out into 'the street'.

Until this moment no one else in the village had been awake or aware of the drama. Suddenly curtains were drawn back, doors opened, dogs began to bark. A goat appeared from the pub car park as the lonely cow careered back towards Haugh House, abluting violently as she went. Trevor Pratt at the Forge, in polka-dot nightshirt and wellies, shouted and waved an umbrella, determined to defend his garden to the death. Gloria Swanson appeared dreamily, dressed in very little, and smiled demurely as Geordie and Tommy ran past.

Thomas Forsythe was slowly driving the pot-bellied pig back to the Pillick menagerie (thinking how close her belly was to the road) when he heard the worried cow and her pursuers approaching from behind. Beyond them he imagined he could see a scantily clad female figure, a goat, and a man in a frock waving an umbrella. The village was definitely going to pieces.

By eight o'clock the community was in turmoil. Only two hours to the judging. Flowers in disarray, footprints on the verges, and a trail of muck all the way up the road. Panic. A brush-and-shovel brigade was quickly organised, a replanting platoon set to work. There was still hope.

At this moment a surprisingly relaxed Polly Pillick appeared carrying a pail full of protein nuts. 'Anyone seen my sheep?' she smiled, 'seems the little rascals have gorn walkabout!' Sheep? No one had seen any sheep. Cattle, yes; a pig, a goat . . . but no sheep. Where the hell were they? Were they lurking somewhere, preparing to leap out in front of the judges' limousine?

The Judges duly arrived at 10.15, and Mrs Graham coffee-ed them as long as she could, while repairs were completed. The gardens were immaculate, they said. Elsewhere some of the flower beds looked a little droopy, – perhaps lack of water . . . maybe a sudden squall overnight, and there were a few mysterious stains on the road. But all in all, very colourful . . . a good effort.

What really persuaded them to put Hindburn onto the shortlist was the churchyard. The old yew hedge was trimmed to perfection, the rose bushes in full bloom, birds singing, bunnies bobbing. And there, lying munching contentedly among the headstones, a picture of perfect pastoral tranquillity, – were fourteen Hebridean ewes and twenty-one lambs. One of the ewes was considerably more tranquil than the others, having choked on a small wreath and expired . . . but none of the gardening experts could tell the difference. It was a sunny day and they were very impressed.

19

In the first week of the summer holidays Gerald sold the house in Acacia Avenue. The purchaser was happy to pay £20,000 less than Gerald had orginally hoped for . . . but Gerald was happy enough to accept. It had to go, – this farming business was hardly a bonanza. Even Prudence was losing some of the naive enthusiasm that had propelled the whole family in the first stages of the adventure.

The children were now bleating that rural life was 'very, very boring'. Any early luke-warm willingness to feed sheep and hens had quickly evaporated, – especially when the task conflicted with 'Neighbours', a Disco in town, or some other extra-mural activity. Transporting the sulking brats to and from 'civilisation' was also very, very boring. Prudence did her best to remain cheerful, but their attitude was disappointing. Things were not working out as she had expected.

Her first disaster was called Margaret. Prudence found Margaret lying very still one lovely June morning, a big single lamb trying to persuade her to get up and provide his breakfast. Prudence immediately summoned Sep.

'She's dead,' said Sep. He could tell as soon as he went through the gate, a hundred yards away.

'What do you mean, she's dead?'

Sep looked at the lady with disbelief. 'The bitch has croaked for God's sake, – she is deceased,' he said, very slowly. 'She has moved to the great pasture in the sky, – she is a late yow . . . she's dead!'

'But she can't be,' protested Prudence. 'She was all right yesterday. Shouldn't we call the vet?'

'There's not a lot he can do at this stage,' Sep assured her.

They were still thirty yards from the body, walking faster, Prudence anxiously flapping her arms about, shouting encouragement. This only persuaded the two crows sitting on the carcase (wondering where to begin their meal) to fly away, expressing annoyance.

'Are you absolutely sure?' she asked, eyes moist, voice unsteady, hope dissolving. 'Maybe she's just got onto her back and can't get up . . . you told me this happens sometimes.'

Sep arrived at Margaret first and kicked the corpse, poked his stick at an unblinking eye. 'I told y' – she's a gonna.'

Prudence was on her knees. 'You're never goin' to give her the kiss of life are y' ?' asked Sep. 'I'm tellin' y', it's a knacker job. They all die sooner or later, – everything does!'

'But she was so happy,' Prudence wailed. 'I'll have to phone Gerald at the office . . . and what will I tell the children, – they'll be devastated.'

Sep was eager to be away. He'd just realised he'd left his cigarettes on the kitchen table. 'Phone the kennels,' he said quietly, 'they'll pick 'er up. Might cost y' a few quid, but what the eye doesn't see the heart winna grieve. I think the lamb can look after himself.'

She didn't even hear him. Still looking at the body, she stood up, slowly took out a crisp white handkerchief with a P embroidered in one corner, and blew her nose extravagantly. 'We'll bury her,' she said decisively. 'Margaret deserves a proper burial . . . I won't have her carted off like some bit of refuse. She's family!'

'Oh my God,' groaned Sep, looking skywards as if for support. . . . 'It's just a dead sheep; I've had hundreds of them. The more sheep you keep, the more you'll lose. In the winter of '63 we lost fifteen in one night, and we certainly didn't have a bloody funeral service for them!'

'Yes, but you're a proper farmer,' she said. 'Our little flock has been decimated.'

'It's one very old yow,' said Sep patiently.

'Why?' Prudence asked desperately.

'Well,' said Sep, adopting the detached tone of a consultant at the General, '. . . maybe pneumonia . . . it can happen, not much warning . . .' He thought he'd possibly been a little hard on the woman, and added quickly, 'Look, pet, if you're determined to be a real shepherd you'll have t' get used to this sort of thing. You have to recognise when something's not right, see it early. I mean animals canna tell y' when they're feelin' poorly. . . .' Then as an afterthought he added, '. . . come t' think of it, sheep wouldn't tell you, anyway!'

'Will you bury her for me?' she asked.

By the time the children came home, Willie had done the deed and Prudence had erected a small cross, with the words, 'In memory of Margaret, a lost sheep'.

She was surprised how 'brave' the children were, almost indifferent. Julian wanted to dig up the remains, remove Margaret's teeth and eyes, – put them in a jar in his bedroom. (Sep could have told him she had no teeth, and the crows had already nicked one eye.) Emma stood by the grave for five minutes crossing herself, and wailing like she'd seen Middle Eastern women behaving on the telly. When she came back she said, 'Can I have a pony now. Ponies hardly ever die. . . .'

Supper was a subdued affair. 'How was school?' Mother asked, trying to sound bright.

'Boring,' they said in unison.

Later, she said, 'Interesting day at the office, darling?'

Gerald despatched a mouthfull of moussaka (Prudence was half-way through Edith Entwhistle's *Ethnic Cookbook*: paella tomorrow, a vegetarian quiche on Friday). 'Not really,' he mumbled, 'business is virtually non-existent. Somebody flogged his Northern Electric allocation, a couple of Tessas. Not much else . . . very depressing. . . .'

'Oh good,' said Prudence cheerfully.

'What about my pony?' asked Emma.

'Haven't you some homework to do?' snapped Father. 'Some maths, French verbs?'

'Quelle tragédie!' said Emma, rising dramatically from her chair. 'Le pauvre moûton de ma mère est mort. . . .'

A few days later there was a cockerel crisis. Prudence had acquired a dozen hens and a cock she named Rasputin, not least because several attempts on his life had failed. The postman ran him over twice, and the bin men almost every Thursday as he attacked their wagon. But each time he just got up, shook himself, and continued his active life in the harem. Tyne, the collie obtained from Willie Turnbull, had ruffled Rasputin's feathers the first time they met, and most days since. Eventually, while Sep and Prudence were discussing business over coffee one morning, Sweep and Tyne combined to mortally wound the cantankerous bird. He fought bravely but was seriously outgunned . . . and ended up in a sorry state.

Prudence was distraught, and rushed into the house for the first-aid kit. Before she could return, carrying rolls of elastoplast and T.C.P., Sep had pulled Rasputin's neck and put him out of his misery. The lady was shocked. 'He died,' said Sep. 'Never recovered . . . if I was you I'd cook him for Gerald's supper, – he'll taste delicious.'

Prudence protested of course, sobbing something about heartless, barbaric peasants. Only after Sep had plucked and gutted the old bird, followed by a week's 'lying in state' in the fridge, did she eventually summon up enough courage to roast oven-ready Rasputin for Saturday night's supper.

'This chicken is quite superb,' said Gerald, gnawing on a leg. 'Where did you get it?'

'It's scrumptious,' said Emma.

'Really brill,' declared Julian.

'Aren't you having any, darling?' Gerald asked, still with his mouth full.

'How could I?' wailed Prudence. 'He was one of us!'

20

Selling the first lambs wasn't easy. It had nothing to do with supply and demand, subsidies, or the price per kilo. Prudence just didn't want to part with them.

She delayed for weeks, conjuring up excuses and a host of other tasks requiring her urgent attention. Sep told her the lambs would become overweight, too fat; prices would fall. Even Polly Pillick had sold several Hebridean sheep and six Vietnamese piglets from the second litter. Even the Peabodys and the Pratts had been persuaded to market some animals and ease the pressure on their paddocks.

Eventually Gerald was forced to call a family summit. They had to get some cash in soon, or Emma would certainly remain horseless, Julian computerless, and Prudence would possibly become farmless. All house renovations would have to cease. There would be no new car this year. He used forceful language, punctuated with a lot of 'darlings', but left no one in any doubt that the position was desperate. Larry, Matilda, Roger and all the other fat lambs would have to go to market!

Willie handled the first consignment. He drew them out,

weighed them, took them to the mart in his trailer . . . saw them graded and auctioned. He handed over a fiver luck money to a buyer who stuffed it into his pocket, and smiled unconvincingly while bidding for the next lot.

However, it was a breakthrough of sorts. A few days later, Sep declared six cattle ready for slaughter. More lambs and one or two bullocks followed almost every week. Prudence bade them a tearful farewell each time. But it wasn't quite so traumatic now, – especially when the cheques began to arrive. Gerald seemed more relaxed.

After a particularly good price for three Charolais, he cut the mart report from the local paper and pinned it on his office wall. 'Northern Auctions sold 212 cattle this week,' it read, 'bullocks to 115p per kilo from Hindhope. . . .' Every client who came in was obliged to read it. Sep told him not to get carried away. 'You could top the mart every week and still end up broke,' he warned.

Came the harvesting of the West field, and another reminder that Prudence had a lot to learn. It had looked much simpler from Acacia Avenue.

'When do we combine the barley?' she asked her adviser.

'When it's ripe,' said Sep.

'But everyone else is cutting,' she protested. 'You have several fields harvested already. . . .'

'Another week,' he said.

'You mean next Monday?'

'Probably, if we can get the combine, and a wagon organised to take it straight to the drier. I'll fix it for y'. . . .'

'But I'm having my hair done on Monday,' she said.

'Cancel it,' said Sep . . . and went away to bale straw at Clartiehole.

As it happened Prudence could have had a perm, visited her dentist, enjoyed a massage and manicure and gone for a leisurely lunch with her best friend, because the contractor didn't arrive on Monday as promised. He didn't turn up on Tuesday either.

'Where is he?' she pleaded on the phone.

Mrs Rogerson, the contractor's wife, was used to telling 'porkies' . . . she was very good at it. 'The combine broke down,' she might say, 'quite a problem getting spares . . . he'll be there as soon as he can. . . .' Or, 'Didn't it rain with you yesterday? Our lads were stopped at two o'clock . . . terrific thunderstorm . . . must've been very local, you're lucky you didn't get it. . . .' This time she said, 'He'll definitely be at your place in the morning.'

Sep said not to worry, the barley was a little harder every hour . . . the forecast was settled. 'Don't panic, woman,' he told her.

By the time Rogerson's big red monster cut the first swath on Wednesday afternoon there was a strong south-westerly wind blowing, and the pickle was coming off at under twenty per cent . . . straight into two twenty-ton trailers. Prudence watched in wonderment as the golden grain poured out from the spout.

The barrister's wife in No.2 Hindhope Court was less enchanted, – as she watched her washing covered in dust every time the machine came down the east side of the field. 'We'll definitely sue,' she declared angrily, 'and an urgent report will go to the Department of the Environment. . . .' She began to cough excessively as Rogerson went past again, enveloped in a cloud. 'And our access road is being destroyed by these massive grain wagons . . .' she said.

She was still gurgling on about pollution and quality of life, the noise, and diesel fumes, when Sep (who'd just come into the field to check that all was going well) told her to 'Bugger off!' 'You can't expect the world to grind to a halt while your knickers dry on the line,' he growled. 'It could rain tomorrow!' With that he jumped onto the step as the combine trundled past, and left Prudence to seek a diplomatic solution. He hadn't made it any easier. They didn't dare burn the straw.

It was a good harvest, helped by long sunny spells that cut down drying costs. Sep declared it 'a canny harvest'. 'No records y' understand,' he said (just in case providence was listening, or someone got the wrong impression), '. . . but we might end up with a wee profit . . . if we're careful. . . .'

Sep was not given to euphoric outbursts.

By the first mellow days of autumn he felt he'd piloted Prudence through her first stormy seas. Without him, Hindhope would've been a boat without a rudder, and everybody knew it. Not that she and Gerald had made any money . . . far from it. The boat was only just afloat. The lamb trade had been 'quiet'. 'Just too many sheep about,' said Sep. All but seven of the cattle had been graded, and they would do all right in the store (somebody else would make a useful profit on them by Christmas). However, there was haulage to be deducted, a couple of visits from the vet and mart commission. With that, and the cost of borrowed money ('our gearing,' as Gerald called it) nobody was growing rich.

And Prudence was slowly coming to terms with what she called 'the calculated callousness' of farming. It took her all year to realise that Sep's apparent heartlessness in times of trouble, especially when animals were involved, was really an in-bred detachment designed to cope with whatever the problem might be. She would never have it.

The death of Margaret the mule yow was her first lesson. Another ewe who expired on her back the day before she was due to be clipped caused considerable anguish. A very good twenty-kilo lamb dropped dead in the pens seconds after being marked for the mart. 'He obviously didn't want to go,' said Sep.

At one stage, all the lambs in the Camp Hill field developed what Prudence called 'severe diarrhoea'. Sep gave her a lecture on various parasitic worms, which left her feeling unwell, and then helped her to dose everything. A third ewe just went downhill in spite of all known medication, until Julian said it looked like Ghandi. Ghandi, however, seemed reluctant to die, and attracted the attention of Mrs Beeday, the plumber's wife in No. 3, who called the RSPCA. They insisted the unfortunate creature be humanely despatched without delay. Prudence was miserable for days.

She had also invested in six more hens to lay dark brown eggs for Gerald's breakfast, together with a replacement cock christened Napoleon. It was a great disappointment when Tyne, pretending to practise his gathering techniques, callously slaughtered two hens and left Napoleon virtually naked. Prudence gave the dog a very serious talking to, – but he got one more before she fenced them in.

The spaning of the lambs caused her more grief than the sheep. In spite of Sep's patient explanations, she found the 'heartless separation' difficult to accept. For the first two days she wandered about, comforting bleating lambs in Camp Hill and the mournful mothers answering from the Lambing field. Residents of the Court complained they couldn't sleep, muttered darkly about 'cruelty in the countryside', and wrote to the Parish Council demanding the lambs should be silent at least until 8.00 a.m. on weekdays . . . 9.00 a.m. on Sundays. By day three it was all quiet, and everyone felt better.

Marts were either an occasion for despair or delight. Despair was parting with 'pets' like Bernie the bullock or Larry the lamb. Delight was buying replacement ewes and tups.

Her first visit to the mart to sell lambs was an acute embarrassment for Sep. She turned up in tight sweater and short skirt, with bright red fingernails and high-heel boots, looking like a million ECUs, and she followed him everywhere. The old regulars sniggered like teenagers. Charlie Sisterson, a spritely seventy-five-year-old with no teeth, invited her for lunch in the canteen. 'Ayatollah' the grader, who seldom smiled at anybody in case it was misconstrued as some kind of favouritism, beamed and grinned, and only knocked one kilo off the half-weight. The butchers even bid against each other while staring at her bosom, – and made no mention of luck money as she shimmied from the ring. Arthur the auctioneer almost fell out of his box to take her hand, and invite her to bring more lambs next week. There was a delay while he coughed and blew his nose, and pulled himself together. Even then his voice wasn't right, and he knocked the next lot down far too quickly.

On the way home Sep gave her a firm lecture. 'Y' can't turn up lookin' like Marilyn Monroe. . . .' he said. (She was the only film star he could think of at the time.) 'The lads lose their concentration when somethin' like that walks into a mart . . . it can ruin the trade!'

By the time they went back to buy at the ewe sales Prudence was a shapeless bundle in duffle coat, wellies and shining morning face.

'Try for these,' whispered Sep, as a pen of useful two-crop mules came into the ring, '. . . they're just what y' need. Nowt fancy, but they've done it all before, and they'll shift nicely. Go on, give him a bid.'

The sale immediately came to a halt as Prudence leapt up and down waving her catalogue in the air. Even the sheep stopped and stared at her, – before fleeing to the other side of the ring.

'What the hell are y' doin', woman?' snarled Sep, sinking deep into his collar.

'I'm only trying to attract the auctioneer's attention,' she said. 'He might miss me. . . .'

'You're jokin'.' Sep was trying to be calm and invisible. 'He'll see you all right . . . all y' have to do is look at the man, nod your head, move a finger. Wink at him.'

'I certainly have no intention of winking at him!' she snorted indignantly. 'He might get entirely the wrong impression.'

'Don't be so bloody conceited,' said Sep. 'All he's interested in is selling the sheep for as much as he can get. If he discovers a mad bidder like you flailing about as if you were drownin', he'll be thrilled t' bits, – y' could end up buyin' the whole mart!'

After that Prudence settled down, hardly breathing, moving nothing . . . only giving a cautious, sideways glance towards the box whenever Sep prodded her gently in the ribs. 'They're plenty,' he would mutter, 'don't give him any more. I think he's taken y' twice anyway. . . .'

'You're out on the left,' shouted Arthur. 'Are you bidding? Last chance. . . .'

'Take no notice of 'im,' says Sep. 'Ignore 'im.'

Three more lots came and went, before another prod. 'Go on,' he said, 'give him a quid . . . but take your time. . . .'

Arthur took her bid, another from somewhere else . . . back to Prudence again. He seemed to hang on for ages before the hammer finally fell, – and she had her first purchase. She almost squealed with excitement. Intoxicated, power crazy, – it was almost like her first serious snogging session in the back of a Morris Minor twenty years ago. Sep had trouble keeping her quiet for the rest of the day. Uncontrolled, she might have bought everything.

In the end she settled for thirty-five three-crop ewes and two cross-Suffolk tups, – and went home walking on air, eager to show them off to Gerald and the children.

Gerald, who'd had an unhappy day at the office with a disillusioned shareholder, only wanted to know what the animals had cost. 'Good Lord!' he exclaimed, without knowing whether or not they were a bargain.

Julian and Emma preferred to ask searching questions about the sex life of the sheep. They giggled a lot, and inspected the nether regions of the two rams. Prudence wondered if this might be a suitable opportunity to discuss the facts of life (something she'd been putting off for some time) . . . but when Sep arrived with harnesses and coloured paint blocks, she abandoned the idea again.

'What would you do without this man?' asked Gerald. 'We should invite him to dinner, I think . . . show our appreciation. I mean, let's face it, darling, if it wasn't for Sep the whole show would be a shambles. . . .'

'Oh I don't know about that,' said Prudence, slightly miffed, 'I'm learning very quickly, darling. . . .'

Gerald looked unconvinced. 'We don't seem to be

making any money,' he said. 'What about the cheque for the corn; has it arrived yet?'

'Oh dear me,' she chirped, 'I almost forgot. It came this morning . . . not quite as big as we thought, I'm afraid.'

'What?' Gerald recognised more bad news. 'I understand we sold thirty-eight tonnes at £110 a tonne . . . that's £4,180! (He'd worked it out weeks ago.)

'Well, not quite,' said Prudence, a little nervously. She paused while Gerald helped himself to a large gin; he suspected he was going to need it.

'Actually, darling,' she said, 'Sep says it yielded quite well, because, as you know, we kept about five tonnes for the sheep, so the field total is well over forty tonnes. . . .'

Gerald would not be moved. 'Thirty-eight tonnes at one hundred and ten pounds a tonne comes to four thousand one hundred and eighty pounds!' he said.

'Right,' said Prudence. 'But by the time it was dried there were thirty-four and a half tonnes. And then there was haulage, and the levy, and a deduction for admix. . . .'

'Admix?'

'Well, a few things that shouldn't have been there,' she explained. 'Some weeds, a little wheat. And you remember those beautiful poppies, darling . . . how delighted we were when the field looked like a Van Gogh landscape, – well, they shouldn't have been there either. . . .'

'How much?' asked Gerald.

'£3,275,' Prudence said, handing over the cheque.

'Oh my God!' Gerald was at the gin again. 'How on earth does anyone make money at this ridiculous business?'

Prudence decided she wouldn't bother him with the contractor's bill . . . not for a few days.

21

I.B.P. launched their bid for Applepeel Electronics on November 1st, when the shares stood at 110p. Sir Humphrey Handyside, chairman of I.B.P., announcing the bid, assured the *Financial Times* that such a takeover was in everyone's best interests. 'Applepeel products fit very well into our long-term strategy,' he said, 'and will benefit hugely from our worldwide network. . . .' He considered his company's offer of 115p to be generous, and was confident it would be accepted by a majority of Applepeel shareholders.

Six months earlier Polly Pillick had asked her husband if trading on insider information was now a criminal offence.

'Probably,' said Giles, 'but there doesn't seem much point trading without it.' He'd just bought another 250,000 Applepeel shares at 83p, all in Polly's name, bringing the family's holding to over three and a half million shares.

The market took note of the chairman's apparent confidence and, after a favourable broker's report, the price rose 7p.

The chairman's confidence was based not so much on

a bulging order book, but rather on an interesting lunch with the aforementioned broker. He advised that I.B.P. had been steadily accumulating Applepeel equity for some time, – and (he suspected) were likely to mount a bid before the end of the year.

A subsequent call to an old school chum in the city confirmed the suspicion, and after arranging loans from the bank and his own company's pension fund, Giles topped up the Pillick holding to five million shares. This helped push the price to 95p. He could have taken a nice profit there and then, of course, but he knew there was more to come. I.B.P.'s buying continued, – and Applepeel topped the 100p mark in October.

When the bid became public they shot up to 125p within a week. Giles promptly sold his entire holding, instructed 'Capability Cleghorn' to look after the menagerie, and flew first class with Polly to the Seychelles.

'Where did y' say they're away to?' asked Sep.

'Seychelles,' said Prudence. She was sitting on the gate watching a tup do what tups are supposed to do. 'Will he make love to all forty-five of them?' she wanted to know.

'I'm not sure *love* comes into it,' said Sep. 'Where's Seychelles?'

'In the Indian Ocean. . . . He seems to have . . . er . . .' (she was searching for an acceptable word) '. . . he seems to have . . . *met* about half of them already. . . .'

'Uh uh,' Sep grunted. 'But he might not have stopped them all. They tend t' leap onto everything for a while, before they settle down t' the job. . . . You ever been to Seychelles?' he wondered.

'No,' exclaimed Prudence, 'couldn't afford it. . . . How long will it take him to . . . er . . . *stop* all of them?'

'Oh just two or three weeks,' said Sep, 'if he's any good.'

'Wow!' said Prudence. 'He doesn't mess about, does he?'

'Not a lot of chattin' up,' declared Sep.

168

Prudence was quiet for a while, watching Rambo, top lip curled, stalking a mischievous mule. 'Have you ever known a gay tup?' she asked.

'Well, you sometimes get a useless one,' he chuckled, 'but not a lot of queers. They're generally all triers. . . . Is it warm in Seychelles?'

'Beautiful,' said Prudence, 'palm trees, golden sands, blue seas. . . . When will we get our first lambs?'

'147 days,' said Sep, moving off. 'April 1st next year, – but chances are some bitch'll surprise y' before that!'

*   *   *

Geordie Dodds changed the raddle to blue on his tup in the Marleyburn field, and stood leaning on his crook, counting red bums. Spot was flat to the ground, eyes bright, making sure nothing wandered off. Geordie noted at least two very lame yowes who certainly wouldn't wander very far . . . at least not quickly.

He counted them all, and strolled up towards West Cottage. Mrs Pratt at the Forge was feeding soft-centred Milk Tray to her livestock: Gareth and Barry, the Welsh cobs, and the Jacob ewes. She'd rented a pedigree tup from somewhere. 'A noble-lookin' beast,' Geordie had to admit. It appeared young Amanda Pratt had more or less abandoned the ponies now, to concentrate on boys . . . Mother was doing all the work. It's usually the way it goes, he thought.

Edith Little was pottering about in her garden at the 'Old Shop'. She spent every daylight hour in there. 'Morning,' said Geordie. And Edith, ready for an excuse to stop work for a while, stuck her fork in the soil and came to pass the time of day.

'Y' keep a canny garden,' said Geordie, casting his eyes over the fruit bushes to the rest of her weedless plot. Cabbages, sprouts and cauliflowers, still green . . . the potato ground dug over . . . a load of muck from his own hemmel steaming gently in the cool air. She was putting the garden 'to bed' for the winter . . . the deep-freeze already full of goodies. Taties and swedes pitted.

She and Fred had most of Gloria Swanson's land as well. Gloria only needed a wee patch of lawn, just enough to sunbathe on. Tommy Cleghorn could tell y' all about that . . . He trimmed her privet hedge four or five times a year . . . always on fine days!

'Y' certainly produce a lot o' stuff from your patch,' Geordie said: 'strawberries, blackcurrants, rhubarb. Fred's won the Leek Show again this year, hasn't he? Aye, it's impressive, – you've both got green fingers right enough. . . .'

'We do our best,' smiled Edith modestly.' Keeps us active, helps supplement the pension. . . .' She took off her gloves. 'Actually, we feed ourselves almost entirely from the garden now,' she said; 'only buy a little skimmed milk, and eggs and meusli.'

'Y' haven't gone vegetarian, have y'? asked Geordie, just a touch of alarm in his voice.

'Oh yes, indeed,' declared Edith. 'We've gone absolutely veggie . . . have been for years, – never touch meat at all now!'

He turned to look at his sheep. 'Y' mean, y' never fancy a nice bit o' lamb with mint sauce?'

'Good heavens, no,' she smiled. It was a very understanding smile. 'We've realised meat is simply not good for you . . . all the recent research confirms this. We're both convinced we'll live a much longer, happier life without it. No cholesterol problems, no heart attacks. Yes, we're devout vegetarians . . . so much healthier, don't you agree?'

Geordie looked at her. 'Well,' he said, 'I'm not sure about that, I'm no expert y' understand. All I know is I've shepherded a few thousand sheep in m' time, and a hellova lot o' them dropped down dead for no good reason that I could ever see. And I have t' tell y',' said Geordie, – 'they were *all* vegetarians!'

He called his dog, and leaving Edith to ponder on that, he set off home for his dinner. Being Monday the missus would have some nice cold beef ready for him. Y' couldn't beat it.

He was walking back through the sheep pens when he heard the screams from Paddock House. Spot heard them too, and looked up at his master as if to say, 'what was that, then?'

Geordie, who hadn't broken into a trot since a deranged Galloway with a new-born calf had had a go at him in the winter of '76, ambled over the road and found little Mrs Peabody in serious trouble. 'Stupid woman,' muttered Geordie.

There, dressed in pale blue overalls with a white lamb embroidered on the bib, pink wellies and woolly ski hat, Diana was trying to put some ewe nuts into the trough for her six mules. Not an impossible task you might think, but the sheep had become so excited at the prospect of nosh that they were continually barging in and knocking her down. The situation wasn't helped by the arrival of the three heifers, who naturally fancied a feed as well.

As he watched, she tried again . . . and, as soon as she lowered the bag, she was overwhelmed and thumped aside. A desperate battle ensued for the one protein nut that had rattled into the trough.

Again she tried (Geordie had to concede she was game). She was covered in clarts, tears running down her face, sobbing obscenities in a posh voice. 'You barstards!' she cried. 'Bugger orf!'

This time she decided to distribute the food at speed. She began her approach fully twenty yards from the troughs, hoping presumably to beat them to it. She wasn't quick enough. Old mule yowes, who half an hour ago might have seemed unable to put one decaying foot past the other, hobbling about with droopy lugs and runny noses, can nevertheless move at the speed of light when offered food, – especially if pursued by a following wind and three greedy, little badly bred heifers.

To be fair, the lady almost made it. However, just as she slowed a little to pour the nuts from the bag, one of

171

the sheep nipped round behind her to get the first mouthful . . . and stepped straight inside one of the pink wellies.

Diana was a gonna. The bag was emptied, the nuts consumed, the trough licked clean . . . and the animals had wandered off before Geordie could rescue her.

'I'll show y' how to do it tomorrow,' he said sympathetically. 'And get yourself a good stick . . . it'll be no bother. . . .'

She was still sitting on the ground with a cold wet bottom, one pink wellie and one soggy green sock.

'Tomorrow,' she said calmly, 'I may instruct Donald to shoot the whole bloody lot!'

22

There were horses dancing in the village. Highly strung, turbo-charged, four-legged Nureyevs and Markovas, entrechating all over the place. Traffic was at a standstill.

The White Hart was doing a roaring trade. 'Are you writing all this down?' asked Nora, as she hurried out with another tray full of shorts. 'Definitely,' said Jack. 'It's all on the Nickers account . . . plus service charge.'

A florid Major-General leaned down from his mount as it careered across the pub car park, and with the grace of a Cossack lancer, effortlessly swept up a large brandy from Nora's tray. The horse was sweating profusely, foaming at the bit. Without warning, it would leap in the air, or rear up on hind legs . . . but the General never spilled a drop. He might have been relaxing on a sofa, instead of perched above the tumult on his self-propelled disaster. He even lit a cigar while the beast was in mid-air, back arched like a banana, – and gallantly doffed his hat to Lady Daphne as she took evasive action. 'Morning, Madam,' he roared; 'splendid day,' – and galloped off sideways up the road.

Apprehensive car owners watched as their Volvos were threatened, petrified mothers desperately clutched their offspring, nervous labradors were put on a lead.

'What a superb seat that man has,' said Lady Daphne. 'Quite superb.'

Hounds were sniffing and pee-ing on anything stationary. The nubile Camilla Forbes-Townsend was attracting much attention. She always did . . . ever since the halter on her Bruce Oldfield evening dress had snapped halfway through the polka at last year's Hunt Ball. On that occasion, Hughie Davenporte had gallantly clasped the distressed damsel to his manly chest, and waltzed her straight off the dance floor into the back of his Range Rover. Today her jodhpurs were so tight, some wondered if blood would ever reach her toes.

Matty Robinson seldom missed a meet. If he didn't turn up, only two possible reasons sprang to mind. Either both his mounts were lame, or Matty had expired. It was a constant source of wonderment how the man could afford such a lifestyle. He had a fifty-acre smallholding carrying two lean tups, a posse of vintage Swaledales, perhaps a dozen pot-bellied stirks, a pack of assorted dogs, a few featherless old hens, an ageing sow . . . and two superb hunters. The horses were fed on the very best hay, while the cattle and sheep coughed their way through the mouldy stuff. Simon and Garfunkel got the sweetest, cleanest oats, – while the hens and pigs gratefully ate the 'sweepin's up'.

Matty himself was turned out like a Lord . . . boots spotless, hacking jacket made to measure, checked waistcoat, white cravat held with silver pin, hard hat at a jaunty angle, and a flask of whisky mixed with cherry brandy tucked away in an inside pocket. On hunting days he also wore a well-practised aristocratic accent, which enabled him to charm the upper crust, and generally reinforce an impression that this dashing figure could be the rightful heir to some vast Scottish grouse moor, rather than a destitute peasant on a mortgaged mount.

174

He took his second drink from Nora's tray with a disarming smile. 'Lovely,' he beamed, and hurriedly reversed Garfunkel between two horse boxes, as the Major-General roared back into the throng shouting, 'morning, morning,' to everyone as they ran for cover, or leapt over a wall.

The widow of Brigadier Hardcastle looked like a well-bred vulture. Perched side-saddle on her black mare, she was (as always) attired in black coat and black topper. Her wizened face was concealed by a dark veil which was only ever raised to allow a glass or a cigarette to her lips. . . then it was quickly lowered, and very few people complained about that.

By contrast, Archie Harrison was a mess in an old sports jacket (tied round the middle with baler twine), dirty wellies and a cloth cap. His horse was a mess, too, often still covered in clarts from the last outing. Archie was a farmer who simply enjoyed riding across his neighbour's land to see how they were getting on. A dead yow lying behind a wood naturally filled him with delight; a patchy field of winter wheat left him overjoyed. It wasn't that the man was cynical or evil, – it just helped to reassure him that his own disasters were no worse than anyone else's. Today he sat easily on his scruffy cob, the reins lying loosely over its neck, while he discussed the week's lamb prices with another hunting peasant. Archie's horse was eating a bush from the Pratts' garden.

The Honourable Nigel Nicholas M.F.H. had dismounted and led his thoroughbred beast among the foot followers, smiling warmly. 'Much better to be on the same level as these people,' he'd told Lady Daphne earlier. 'One can give entirely the wrong impression talking down from seventeen hands, what!'

'Quite,' she said. 'In these times of anti-blood sport fanatics, we need all the friends we can get dear, – even if they're just ordinary people. . . .'

An ordinary person in a yellow and purple anorak touched the spot where his forelock had once been, and asked if his Lordship thought he'd get a good 'run' today. . . .

'Oh absolutely, tremendous fun,' bellowed Nickers, and moved on among his 'subjects', lifting his hat to the ladies, trying desperately to remember names.

He'd just paused to greet Samuel Forsythe when the Protesters emerged from the White Hart.

The five young men and a girl, all dressed in camouflage gear, big boots and long woolly scarves, were wielding banners and shouting 'save the fox' and 'stop the hunt'. Some of their other outbursts suggested all those on horseback had been born out of wedlock. One of the saboteurs began sprinkling pepper over the hounds, another lit firecrackers and threw them into the mêlée. Dogs sneezed, horses bolted off in all directions, petrified ponies with red ribbons on their tails kicked each other enthusiastically. A small Right Honourable on holiday from Harrow was deposited in a ditch, and began to wail. A demented gelding, only recently persuaded from his horsebox, tried desperately to get back in, – even though the tailboard was now up.

The General, who'd been galloping backwards round and round the war memorial saluting, returned to take command immediately.

'Cry God for Nickers, England and St George,' he roared, 'Gentlemen now a-bed shall think themselves accursed they were not here!' Downing his brandy, wielding his whip as a sabre, he charged the enemy with such ferocity that they fled in terror, – scrambling into an old dirty white Ford van, strategically placed for the getaway.

Nickers had remounted and was trying to persuade his horse to destroy the van. Others had joined the battle, and were beating the roof with their whips. It seemed to take an age to start the machine before it stuttered off in a cloud of blue smoke heading towards Clartiehole, pursued by the cavalry and cheers of approval from the Barboured infantry.

It was almost half an hour before peace was restored. Ferrett the huntsman gathered up his spluttering pack, blew his horn, and led the victorious army away to draw the wood on the south road. Fanatical followers climbed quickly into their cars, while the less enthusiastic retired to the bar. Jack and Nora totted up the bill for Nickers, and the village was left empty again, save for a lot of little brown heaps, steaming on the road.

The hounds found little to excite them in the wood. A brace of pheasants flew out, protesting noisily, and zoomed low over a cluster of temperamental horses who, assuming these things must be nuclear missiles, took off in blind panic. Camilla Forbes-Townsend's mount, which had never jumped anything higher than a thistle in its life, tried to leap over a five-barred gate, but only succeeded in destroying it completely. The sinister widow Hardcastle found herself transported at speed towards the Marley-burn, where her horse braked violently. Mrs Hardcastle carried on alone, but failed to clear the water.

Other more disciplined groups stood patiently in pre-pared positions around the wood, waiting for Mr Fox to appear and play the game . . . but he wasn't at home. Ferrett called the pack together, and moved over the road into a field of turnips.

Meanwhile, six people in a smoke-filled, off-white Ford van were drinking lager and wondering what to do next.

Wayne (second-year sociology, Gateshead Polytechnic), self-appointed leader, acne lingering among the ginger stubble, dew-drop on nose, crouched menacingly behind the wheel, revving the engine. He dare not switch it off in case of another sudden attack. Sandra peered anxiously out of the passenger side window, searching for the enemy. The other lads were huddled uncomfortably in the back.

'Might as well go home,' suggested Kevin. 'We've done the demo, and I don't fancy being kicked to death by them crazy horses . . . they're big 'n' nasty things, those horses. . . .'

'Like, massive,' said another voice from the gloom in the back.'

'It's not the horses' fault,' protested Sandra; 'it's those toffee-nosed sods who sit on them. They're murderers!'

'Barbarians,' said the voice.

'There they are!' shouted Sandra, suddenly. 'See them galloping across that hill over there . . . hundreds of them!' She wound down the window, and the cries of the hounds and wail of the horn blew in. 'They must've found a fox,' she said. 'They'll slaughter the poor thing and paint themselves all over with the blood! C'mon, we've got to save him.'

'Let's go home,' moaned Kevin. 'I'm bloody freezing t'death, and real hungry. . . .'

'Wayne!' Sandra was beating her leader about the head. 'Go, man, go!' she yelled.

'Like Action Man,' said the voice in the back, wearily.

Action man needed both hands to grind the machinery into bottom gear before they could move off towards the fray. 'We'll park at the top of the hill,' he said. 'Then, when those bastards chase the fox over the road, we'll rush down and block that gate with the van and stop the horses . . . right?'

'We'll all be killed,' bleated Kevin. 'They'll gallop all over us . . . we'll be trampled to death, they'll show no mercy. . . .'

'Massacred,' said the voice; 'like Custer's last stand.'

At the top of the hill the four in the back got out to breathe a little fresh air and urinate lager into the hedge. They could see the Hunt in full flight, dogs in full cry, coming over the horizon . . . flashes of red, yells of excitement, thundering hooves. The charge of the 'Tight Brigade'.

'C'mon, let's go!' cried Sandra.

'Not yet,' said the bold leader; 'these operations depend on perfect timing y' know.'

Kevin was zipping up his camouflage trousers when he saw the danger. For a second or two he couldn't believe it. He looked over towards the hounds, streaming and yapping down the field towards the gate, then looked back up the road to the approaching disaster. He turned again and saw the riders led by Nickers and Ferrett pouring over the hill. They were over there; so what was this scarlet thing coming at the van from behind? It had no right to be here.

The Major-General had long since emptied his flask, and was in no mood to be trifled with. When he caught sight of the white van, he recognised the enemy immediately, and charged . . . sideways. 'Into the valley of death rode the gallant six hundred,' he roared, spurring his bewildered steed towards the foe.

'Christ!' exclaimed Kevin, and leapt back into the van. The other three heroes (who had stood open-mouthed and open fly-ed for a brief moment) tumbled in after him, and pulled the door shut.

'Go!' they screamed, 'Go!'

Wayne and Sandra heard the terrifying sound of the General's approach . . . 'Blaggards!' he shouted as the horse leapt, jumped, snorted and zig-zagged towards them. He was standing high up in the stirrups waving his whip. Occasionally it seemed the horse might not be there when he came down again, but somehow the General's superb seat always landed on the saddle.

Thankfully for the saboteurs, the charge took a long time. He'd covered about a hundred yards since Kevin first saw him, none of it in a straight line. However, he was almost at the rear windows now. Two more side steps, another pas de deux, and he'd be upon them!

'For God's sake, go!' they pleaded, as the first whiplash hit the roof. 'Wayne!' they cried in unison, as a hoof thundered into the side. The General was threatening them with hanging. The horse was kicking with both back feet now. The van was becoming smaller!

At last Wayne got the wreck into gear again, and it began bouncing off down the hill, gathering speed, with the General leaping and kicking and screaming behind them, – galloping from one side of the road to the other.

In the dip at the bottom the hounds were scrambling through the fence and crossing the road, noses to the ground, while Nickers was trying to unchain the gate. Riders were piling up behind him, waiting to get through. Some of them were waving whips and swearing.

Wayne didn't stop, he didn't even brake as he tore past the site of what was to have been their last, brave stand. Somehow he missed all the hounds, and the van belched its way up the slope towards the main road and safety. As they reached the crest of the hill, they ran over a fox as it darted out of the hedge. It didn't stand a chance.

23

Harry was one of three postmen who did a turn round Hindburn parish every third week. A uniquely rural collection and delivery service, six days out of seven, come hell or high water.

Sometimes they would only bring an electricity bill or the *Farming News*, or a basketful of worthless technicolour special offers destined for the bin. But if, for whatever reason, they didn't call at all, then for twenty-four hours it was feared the world might have stopped spinning.

Gladys was opening Christmas cards, praying there'd be none from someone she had forgotten, when she came upon the invitation: a simple white card with a wine glass drawn in the top left-hand corner, and the message in flowing script.

She passed it without comment to Sep who held it at arm's length (his specs being elsewhere) and considered it as if it were a summons from the tax inspector.

'Should be interesting,' said Gladys. 'We'll have to get your suit cleaned. . . .'

'Wouldn't bother,' he muttered. 'I think I might have flu that night.'

Gladys ignored the threat. 'It'll be great,' she said. 'We'll meet new, fascinating people. . . .'

'Townies,' he said dismissively.

'Well, we only ever see your old farming cronies, and a right miserable lot they can be. All they talk about is yowes and cows and sows, and swop lies about their barley yields. It can be very boring for us women, y' know. . . .'

'It'll be all collars 'n' ties 'n' posh frocks,' he said. 'White wine and one-bite vol au vents.' He made it sound like a ten-year sentence in a Turkish jail.

'That reminds me,' said Gladys; 'I'll need a new dress. . . .'

Maybe he hadn't heard. 'There'll be nobody to talk to,' he groaned.

'Don't be ridiculous,' she snapped. 'The Pillicks 'll be there; they say he's a millionaire now, y' know. . . .'

'The man's a scoundrel,' said Sep. He was making himself a cup of coffee, while Gladys gave further consideration to the party guest list.

'The Peabodys for sure,' she went on, 'and the folks from the Forge. . . . That lawyer chap from Hindhope, – Prudence will *have* to invite him and his wife. The Grahams from the vicarage, maybe. . . .'

'That lawyer's too clever by half,' he snarled.

'All the new people from the farm, I expect,' she said. 'Some of Gerald's city business friends probably . . . that Gloria Swanson woman.'

'Oh do y' think so?' Sep sounded a little more interested.

'. . . and I'll need a new pair of shoes,' said Gladys.

'Shoes!' exclaimed Sep. 'Shoes! You've got at least three pairs that I know of.' He considered this extravagance for a moment. 'You're the Imelda Marcos of Hindburn, you are,' he growled. Gladys disregarded the jibe, conjuring up a whole new image to compete with the sophisticated townies.

She'd have to be very clever with the housekeeping for a week or two, – that's for sure.

The party (Gerald referred to it as 'our little drinkie-poos') was on the Friday between Christmas and New Year . . . and Sep had a few problems that day.

Willie discovered a burst water pipe near the trough in the hemmel, and they had to move a mountain of soggy muck to find it, while being licked by curious, slavering cattle. An old yow, who'd been threatening to die since September, had recently achieved her lifetime ambition. She'd lain unburied for more than a week. Sep had been meaning to dispose of the carcase for some time, but now, with the Hunt meeting nearby and likely to run over Clartiehole, she would have to be interred before the next day. The remains were now scattered over a wide area, and well past their burial-by date.

In the afternoon Sep and Willie loaded up the silage trailers for the rest of the weekend, and by nightfall they both had a distinctly agricultural aroma about them. Gladys got a whiff of Sep even before he came into the kitchen; a pungent peasant perfume of muck, ripe mutton and fermented grass drifted in from the back porch as he took off his dirty wellies. Even half an hour's soak in a bath liberally supplemented with 'Fragrant Apple Flower Foam', followed by excessive use of aftershave and various deodorants, failed to eliminate the problem completely.

Gladys sniffed as they passed at the bathroom door. . . . 'It's still there,' she said, 'just.'

'It's far worse,' growled Sep. 'I smell like a bloody tart now. If there's any weird blokes at this party I'll be in dead trouble!'

They were late. The car wouldn't start, and Gladys was invited to push it down the road until it fired into life. Then they had to go back again to clean her new shoes.

At Hindhope, Gerald came out from the babble of voices to meet them in the back porch. (Gladys had wanted to go to the front door, but Sep said nobody with any sense ever went to the front door of a farmhouse.

'For a start, they probably wouldn't hear y' knockin',' he said, 'and even if they did, they wouldn't answer anyway. 'They'd reckon you were a Jehovah's Witness or a Safety Inspector!')

Gerald took Gladys's coat and ushered them both towards an array of bottles and 'nibbly things' in the lounge. Prudence waved from the kitchen.

'I've been in this house a thousand times,' said Sep, 'and this is the first time I've got beyond the kitchen. . . .'

Mrs Beeday, the plumber's wife, was pouring herself a large gin. 'Have you recovered from your bad turn at the clippin'?' Sep asked. He remembered her being led away from the bloody scene when Alfie cut a lug off.

The awful memory flooded back. 'Oh yes, my God, it was horrific,' she squealed. 'Absolute carnage. . . . I didn't realise farming could be so . . . so brutal!'

'Good job y' weren't around at the castratin',' grinned Sep. 'I used to pull them out with m' teeth, y' know. . . .' He might have gone on, but Gladys dragged him away.

Trevor Pratt came over. 'Ah, you're just the man I want to see,' he gushed.

He came much too far into Sep's private air space, but possibly got a quick whiff of the lingering farm fragrance, and withdrew again to a more respectable distance. 'We've got a sheep problem,' he smiled.

'Who hasn't?' said Sep, thinking there was nothing much to smile about.

'Poor thing's wandering about with her head on one side, twitching and blinking . . . hasn't eaten for weeks. One might get the impression the creature has some kind of brain damage. . . .'

'That'll be it,' said Sep, topping up his glass.

'Well, you're the expert,' laughed the patronising Pratt. 'What do we do about it?'

'Nowt,' muttered Sep. 'Just bury the bitch when the time comes. . . .' This seemed to dissolve Trevor's smile, and he soon shuffled away to charm Gloria Swanson and talk of more lively things.

184

'You should circulate,' whispered Gladys, passing by . . . and to be honest it suddenly didn't seem such a bad idea. He felt much better, more confident. He might even grab Gloria before the night was out. Was it the whisky?

He barged into a quartet of smartly dressed men all talking at once. 'This chappie will know,' said one of them. Sep recognised the barrister from No.2. 'We were discussing Gerald's farming enterprise, – and got onto this mad beef we're all eating these days,' he grinned.

'What mad beef?' Sep growled.

'Well, we've seen it on the telly,' said Q.C. 'The evidence is indisputable. Hundreds of cattle staggering about on our screens every night . . . quite terrifying . . . appalling. . . !'

'It's the same poor cow every time,' said Sep sharply. 'It's the only bit of film they've got. . . .'

The barrister refused to be dismissed so easily. 'That may be so,' he conceded, 'but haven't they discovered the same damned virus in a pig and a sheep? And Lord knows what else for all we know. . . .'

Sep's empty glass was whisked away and a full one put in his open hand. 'Well,' he said, taking a sip, 'if some over-eager young vet injected you with rabies – just to see if a lawyer would foam at the mouth like a dog – y' might well bite the silly bugger, – don't y' think?'

Gaskett, the car-dealer from No.1 Hindhope Court had been hovering on the fringe. 'Tell me,' he asked, pushing in. 'How does it feel to be paid to produce nothing?'

Sep looked at him. He felt he was being baited. 'No idea,' he said. He didn't want to be dragged into *this* field. He wasn't *that* drunk. However, they weren't going to let him off the hook so easily. 'Come along,' said Gaskett, 'what about all this set-aside nonsense at eighty quid an acre? Didn't I see some fellow had set his whole farm aside, and was raking in about fifty grand a year sitting on his arse?'

Sep had the uncomfortable feeling of a man in the dock. The Q.C. fixed him with his best, withering

185

courtroom stare. 'And half the world starving!' he snarled.

'True enough,' said Sep, as calmly as he could. 'But that's the system. . . . I can't lie awake worrying about Africa every night, – just because the politicians have decided I'm growin' too much, and they don't know what to do with it. . . .' His glass was empty again.

'I certainly can't afford t' grow wheat and then give it away to somebody because *he*'s hungry . . . it wouldn't be long before I was starvin' as well. Aye, it's an unfair world,' he said sadly . . . 'and t' be honest I'm not in favour of the scheme at all. I can't see *m'self* goin' into set-aside . . . wouldn't fancy gettin' up in the morning just t' walk round a wilderness. But y' never know, if I was really hard up – no cash to finance the crop, makin' no money, the bank breathin' down m' neck – I might be forced into it.' He paused to accept a re-fill, and wondered if anybody was listening to him.

'In the meantime,' he went on, 'they reckon it's cheaper to pay some of us to grow nowt than it is to subsidise a surplus piled up in a store somewhere. It won't last for ever,' he said; 'the land'll be needed again some day (what's left of it). Question is, – who's gonna farm it?'

Donald Peabody from Paddock House put his arm around Sep's shoulder. 'We peasants have to stick together, eh?' he laughed. He appeared to think this was a huge joke. Sep wasn't so sure, and quickly slid out from under the embrace. Maybe Peabody was a bit . . . y' know . . . strange?

A dapper little man in a pale blue jacket, yellow trousers and a spotted bow tie said, 'You're a farmer, I believe. How fascinating, always wanted to be a farmer. I do so envy the total tranquillity of it all.'

Sep had a sudden picture of this twerp trying to catch a deranged Cheviot hogg hanging a lamb on a cold wet March night, and wondered how tranquil he'd be after half an hour . . . but Prudence interrupted his thoughts. 'Supper's ready!' she said.

Supper was a buffet affair . . . help yourself from the

spread, and find somewhere to sit and eat it. Sep found himself on the floor in a corner between G.G. Graham from the vicarage and Gloria Swanson. It was not entirely by accident.

'So how's business?' asked Graham cheerfully . . . and then proceeded to tell Sep how his own business was doing. 'Difficult times,' he said, 'tremendous cash flow problems, nobody pays their bills any more. . . .' He seemed able to eat, drink and speak all at the same time. Perhaps that was the secret of a successful businessman. . . .

'Flew to Hong Kong on Monday,' he said, – 'back via Singapore, stopped off in Frankfurt, orders worth two hundred K. We may have to re-tool, – gearing dangerously high already. . . .' The man droned on, apparently convinced it was all fascinating stuff.

'Canny supper eh?' said Sep, turning to smile at Gloria's very low neckline.

'Lovely,' she beamed. 'You're Sep, aren't you?' Prudence has told me all about you. Says you've been a tremendous help, – her guide and mentor. I suppose you're very experienced. . . .'

He wasn't too sure how to answer, so he grinned his boyish, embarrassed grin and said, rather coyly, 'Oh I wouldn't say that, pet. . . .'

'But you've been a farmer all your life, haven't you?' She seemed really, genuinely interested. Her shoulder was touching his. 'Farming must be enormous fun.'

'Fun' wasn't a word Sep used very often. 'Fun?' He considered the word; no, it didn't sound right.

'I wouldn't say it was *fun*,' he said. 'It can be quite rewarding, I suppose. But anyway I couldn't do anything else, – not now.'

'You're not that old, are you?' she smiled invitingly.

'Probably past m' best,' he said, and she laughed. He couldn't help noticing that certain areas of Gloria wobbled gently when she laughed.

'At least you're your own boss, independent,' she went

187

on. 'That must be worth a lot?' Her eyes were wide open, giving him her undivided attention.

He'd consumed three glasses of red wine with the chicken casserole, and this, added to the whisky, was affecting his knees. The right one was beginning to develop a mind of its own. It was rubbing against Gloria's left thigh. He thought he'd better say something.

'Did you read that a cockerel in Ulverston is only allowed to crow between seven and seven-thirty every morning,' he said. 'People were complaining about it, something t' do with the noise abatement laws. Would you believe that?'

'Good heavens,' she said, her face very close to his.

'Oh that's right,' declared Sep. 'Rules and regulations ruinin' the job. It's those foreigners in Brussels y' know. There's a bossy little bureaucrat called Delors over there, – Jacques the phantom frog. Seems determined t' rule the world . . . another bloody Bonaparte,' he said. 'We can hardly keep up with the paperwork now!'

'And of course you have to consider the enviromnent as well, these days,' said Gloria breathlessly.

'Absolutely, pet. Y' can't just do what y' like y' know, there's special rules for every little job: pollution inspectors plodging about all over the place, animal welfare officers watchin' your every move . . . we spend more time keepin' the officials happy than we do farmin', these days!'

'Gracious, I had no idea,' she said, eyebrows raised. 'You must have to be remarkably clever to cope with all the pressure. . . .'

Somebody had sneaked up and given him another glass of wine, and he noticed his right arm was draped over Gloria's leg. 'Tell y' the truth,' he confided, 'I'm hardly clever enough for this modern agriculture now. I mean, take the breedin' side for a start . . . all this frozen embryo stuff. They're turnin' hens into roosters, they've got genetically engineered sheep . . . it's all done in a bloody test tube, y' know!'

'How uncomfortable,' she giggled. 'I'm sure the old-fashioned ways were much better. . . .' His right hand had somehow alighted on her left knee.

'We're going home dear,' said Gladys. 'It's one o'clock, – way past your bed-time. I'll drive.'

As they were leaving, a big white BMW screeched to a halt in the yard, and a young couple in evening dress leapt out, slammed the car doors, and rushed past them towards the house.

'Tarquin, my dear fellow,' cried Gerald from the back door, 'we'd given you up for dead!'

'I'm afraid we got lorst,' said Tarquin desperately. 'Been twice round the damned county looking for your pad . . . we're absolutely exhausted.'

'I expect you need a very large gin,' Gerald smiled, 'come on in. You can stay the night if you like. . . .'

Gladys drove carefully out onto the road. 'Did you have a nice time then dear?' she asked.

'Aye, not bad,' he said. 'Good supper. Hardly spoke t' anybody, though. There was nobody there I knew . . . all townies.'

'Y' seemed t' be getting on well enough with one of them. . . .' She looked sideways at him.

'Tell y' who wasn't there,' said Sep, ignoring the probe: 'the Pillicks from Haugh House, – are they still in Seychelles?'

'No, he's in Brixton,' said Gladys. 'Apparently the fraud squad picked him up at Heathrow last night!'

'Serves 'im right,' was all he said.

24

Everything came to a head on a Saturday morning in April.

It was snowing. The electricity had gone off during the night . . . and Harry the postman brought the other bad news all at once. He seemed to recognise the signs. He didn't hang about. There was obviously a deepening depression in the Hindhope kitchen.

Today another letter from the bank expressing anxiety. Some figures from the accountant confirming worst fears. Final demands for spray, fertiliser and the Kawasaki ATV (with Logic trailer) acquired before Christmas. There was trauma outside as well: two dead ewes and some lambs with E. coli.

Prudence was boiling a pan of water on a camping stove to make coffee. The children had decided there was no point in getting out of bed while both life-support systems (TV and Radio 1) were out of action. Gerald was feeling very low.

'We're in a spot of bother, darling,' he said, surprisingly calm.

'Oh,' said Prudence. 'Is it serious?'

He was staring out of the window towards the West field, where several million pigeons were coming in to land. 'It's not good,' he mumbled. 'At this rate I believe we could be bust by June. . . .'

'We'll have some lambs to sell by then,' she said hopefully.

'I presume you're referring to those that won't actually diarrhoea themselves to death,' he said cynically.

'There's five geld ewes going to the mart next week,' she offered.

'Too little too late,' he said.

'Forty tonnes of corn in August.' Prudence handed him a mug with coffee granules floating on the top. 'It might be a bumper crop.'

'Much too late,' he said. 'Actually, it's been staring us in the face for a while now. Didn't want to admit it, I suppose. There's just no way anybody can make money pussy-footing about on seventy acres; it's not possible.' He came to the kitchen table and sat down wearily. 'We'd be better off doing *nothing* . . . your friend Sep would tell you that, darling!'

'Is there no way out?' she asked, very quietly.

'Afraid not,' he said. 'It's time to bite the bullet.' He made a face as he tasted the coffee, but made no comment.

On Monday the lights were on again; the snow was gone, and so was the Kawasaki. Julian protested that if Emma was to have a horse he should have the bike, but Prudence, almost in tears with embarrassment, returned the machine to Anderson's Store. Joe told her not to worry, it happened all the time, – and sold it the same day to Diana Peabody, for cash. Mrs Peabody had thought for some time it would be very useful for leading hay to her livestock. She was heartily sick of being trampled half to death every time she ventured into the 'ranch', and having to crawl back through the fence covered in muck.

191

Geordie Dodds got his eye on the vehicle, and asked how she came by it. She told him she understood it had been returned from Hindhope. 'A snip,' she said. 'Nearly new.'

Geordie mentioned it casually to Tommy Cleghorn, who in turn passed on the information to Polly Pillick, when he was up there digging her garden. 'Tremendous machines,' he told her, they can go anywhere. You should have one . . . any fool can drive them. . . .'

Polly was suddenly persuaded such an instrument was absolutely essential for the transportation of llama nuts and deceased Hebridean sheep . . . but phoned Prudence first, to ask why she'd parted with it.

'We may have to part with everything soon,' sobbed Prudence.

'Oh dear,' said Polly, and promptly phoned her husband who was out on bail and in conference with his solicitor. 'Hindhope's for sale again,' she chirped gleefully. 'Shall I nip up and buy it?'

'Steady on,' said Giles. 'How did you come by this news?'

'Can't talk now,' said Polly urgently. 'I'm orf to buy a tricycle.'

Giles put the phone down, and repeated this strange conversation to his solicitor. The solicitor told the amusing tale of his client's mad wife at court next day to a barrister friend who lived somewhere near Hindhope. The Q.C. from No.2 swooped at 7.00 pm. He waited until he saw Gerald's car was at home, and knocked on the back door. 'Oh God,' exclaimed Prudence, peering out of the window, 'that's all we need . . . what's he about to sue us for this time, I wonder? His kids probably have E. coli!'

Sep was given the sad news the following weekend. Indeed he'd been half-expecting it for some time. They had to be losing money on the farming of course, he realised that, – but he'd always assumed Gerald was rich enough to stand it. There was even some talk of

Prudence's old Dad being very wealthy . . . but he'd gone and married a bimbo in South America, and she might even have some kids. Well now, if that happened, maybe Prudence wouldn't get very much after all. You can never budget for these things, can you?

'So what are y' going to do?' he asked, leaning on the kitchen bench, cup in hand, cap pushed back to reveal the borderline on his brow twixt sun and shade. A dehydrated lamb shivered in a cardboard box at his feet. It was very smelly. 'You're not packin' in are y'?'

'Well, not entirely,' said Gerald. 'We still want to live here . . . the house is great, but we just can't afford to go on pouring money down the sink. . . .' He paused, arms outstretched, inviting understanding. Sep said nothing.

'The bottom line,' said Gerald, 'is that we can afford to live here all right, but we certainly can't afford to *farm* here!'

'So what do y' do now?' asked Sep.

'We've had an interesting offer from the barrister,' Gerald grinned. 'I think the gang at the Court must have got together. They want to buy Camp Hill, West Haugh and the Lambing field, a field each I imagine, at a very fair price . . . more than we paid for them. It's quite tempting. It would certainly solve a few problems.'

'Do you need that sort of money?' Sep asked.

'Well, naturally it would be very welcome,' Gerald answered, carefully.

'But if you stopped spending on cattle and sprays and all that stuff . . . would that do it?' Sep had taken a chair and sat at the table. Prudence gave him more coffee.

'Perhaps,' said Gerald. 'Probably . . . but we're definitely haemorrhaging . . . and we have to stop it somehow!'

'You could probably let it as grass parks . . . if you don't *have* to sell up,' suggested Sep.

'Oh we've been considering all sorts of alternative schemes for a long time . . . this hasn't just happened overnight, of course. We thought about a herd of goats, perhaps . . . cheese, angora wool, that sort of thing. . . .'

'It's been done t' death,' said Sep. 'Everybody's had a go at that. . . .'

'And reindeer farming, there was an interesting article in *Good Housekeeping* . . . it attracted Prudence.'

'Aye, I don't suppose they're entirely trouble-free either,' Sep muttered.

'Poultry, we thought about them too,' Gerald went on. 'Selling farm fresh eggs at the gate.'

'Y' want very little t' do with hens,' growled Sep. 'A hen is not an intellectual animal. In fact, it's possibly the stupidest thing on a farm . . . sheep second, horses third.' He considered his league table for a moment or two. . . . 'But I suspect a mixture of goats and reindeer might well beat the lot!' he concluded.

At the sink, Prudence was trying to remove the smell of lamb scour from her red hands. She had washed them in Fairy Liquid, Imperial Leather, carbolic soap, T.C.P. and rubbed in Gerald's after-shave, – each time raising her fingers and sniffing. 'Oh God!' she pleaded to no one in particular.

'What about pigs?' Gerald asked.

'What about them?' Sep didn't sound enthusiastic.

'Well we did consider having a lot of sows in huts out in the fields, rearing their litters . . . I'm told they're highly intelligent and very prolific.'

'Pigs,' said Sep, elbows on the table, hands around the coffee cup, '. . . either feast or famine . . . you'd probably lose another bloody fortune!'

'Fish farming?' Gerald carelessly threw the idea onto the table.

'Jesus Christ!' Sep suddenly sat upright, 'Where the hell do you lot get these brainwaves? Do y' think y' just toss a few trout eggs into a pond, and wait for them to grow up and surrender?'

'It was just a thought,' said Gerald weakly.

After a long silence, Prudence (flapping her smelly hands about) said, 'we'd really hoped to be totally organic, you know. . . .'

194

'Y' can't afford it,' said Sep.

'Yes, but in the long term it has to be the only way, – surely,' she persisted. 'We can't go on destroying the planet with chemicals and drugs.'

'Ah, but you haven't got a long term!' Sep wished he hadn't said that . . . it sounded a bit callous. 'Anyway,' he said, 'that's a scenario for a perfect world, – where everybody has plenty to eat, and money to pay for it. . . .'

'I'm sure people will be happy to pay more for really healthy food,' argued Prudence.

'Maybe,' muttered Sep. He had a picture in his mind of a poor crop of potatoes, enveloped by exotic weeds, and bondager women searching desperately for the organic King Edwards.

Julian and Emma had struggled out of bed and switched on the TV. 'Guns and Roses' were beating out a Saturday morning message for the teen scene. It was time for Sep to go.

'Let's know what y' decide,' he said as he left, 'Willie's dosin' lambs this mornin' . . . I should be giving him a hand.'

All the following week Gerald and Prudence considered their dilemma. On Friday night the Q.C. rang on his car phone, while stuck in a traffic jam in Edinburgh. He'd talked with Gaskett and Beeday, he said, and was prepared to offer another three thousand pounds. He wanted a decision quickly.

'What will they do with the land?' asked Prudence.

'More gee-gees I expect,' said Gerald. 'They've already got half a dozen in the croft, and it looks like the Gobi desert. They have enormous appetites, horses.'

'What will they do with the land?' asked Willie, when his father told him the plot.

'More bloody gee-gees I expect,' muttered Sep, 'some sheep maybe. I suppose if that lawyer finds out about the ewe subsidy, they might go mad and have a tup-full each. But I expect they'll just mess about with horses, a goat or two perhaps ... a couple o' hens. .... Y' know what these folks are like – the Pillicks, the Pratts and the Peabody's, – they're all the same ... give them a few acres, a pair of wellies and a hamster, – and they reckon they're farmers!'

'Could we take it on?' Willie wondered. 'Rent it, perhaps; grass park it, maybe?'

'It had crossed m' mind,' said Sep. 'What do y' think, should we make them an offer? Or just hang about for a while, and wait for them *all* to give up. . . ?'

'Well,' said Willie, 'I don't think I'd want to be workin' in amongst that lot every day, anyway. They'd be complainin' about something all the time. . . . The muck midden would be in the wrong place, the silage pit would be leakin', the tractor would be too noisy, the cattle would shit on their patios. It's not worth it, is it? I mean there's four townie families up there now . . . it could be like farmin' in the middle of Gosforth!'

As it happened, they didn't get the chance. Prudence called to tell them the news. She felt that her friend Sep should be the first to know. She looked ten years younger, suddenly very attractive again, as when he'd first seen her. Was it just last spring? She wore a perfume today that certainly didn't remind him of sheep.

'We've sold,' she said excitedly. 'They've got jumps up already. Mrs Gaskett is doing dressage on a hairy Shetland, the children are gymkhana-ing and squealing . . . dogs everywhere! Q.C. and Beeday want to buy some ewes and lambs as well, – we thought you might take the rest. . . .'

'If I can have m' pick first,' said Sep.

'Absolutely,' said Prudence. 'Naturally . . . and we're keeping the West field, – sow it with grass (she pronounced it grarss) after the harvest . . . perhaps you'd like to rent it?'

'Perhaps,' said Sep, looking cautiously over towards Willie, – both of them imagining the complete chaos at Hindhope. The upwardly mobile commune indulging their rural fantasies. 'We'll see,' he said.

'It's all very exciting,' said Prudence. She was glowing. 'Gerald's delighted, of course. He never really wanted to farm in the first place. To be honest I probably wasn't cut out for it, either . . . just seemed such a soopah idea, a lovely way of life. . . .' she smiled. 'Not as easy as I thought, though. . . .' (Sep remembered Margaret, her first dead yow, and grinned.)

'Anyway it's certainly taken the pressure off,' she said. 'It was all becoming rather fraught, as you know. Now we'll be able to finish the house as we'd planned; we may send the children away to school, Emma can have her pony. Gerald's already bought a new car from Mr Gaskett, – a Mercedes I think. And the neighbours won't threaten us any more (she laughed), – we're all in the same boat now, more or less. And of course we'll have that beautiful house . . . Mr Cleghorn's busy in the garden now. (Sep couldn't recall Tommy having been referred to as *Mr* Cleghorn before.) So if you'd like to come and choose your sheep in the morning, that would be lovely. . . .'

'I'll want young 'n's wi' twins,' said Sep firmly; 'and they'll have t' be a reasonable price. . . .'

'Of course,' she said. 'In fact I'll need you to tell me what to charge Q.C. and company for theirs!'

Sep watched her slide into the car, smile and wave, and roar away in top gear with the clutch half-way out.

'Canny woman,' he said quietly.

Willie took a deep breath, and hitched himself up onto the gate at the sheep pens. 'So that's it,' he said, 'the end of the farm. Hindhope's gone . . . just a playground now.'

'Aye, it's a pity,' muttered Sep thoughtfully. 'Funny how the picture changes. . . .' He pushed his cap back to scratch the top of his head. 'When old Forsythe first went there, y' know, – nobody wanted the land at any price . . . couldn't give it away. The war changed all that, I suppose. Now everybody wants a bit, but nobody wants t' farm it . . . 'cept silly buggers like you 'n' me . . .' he grinned.

Also published by Farming Press Books

Country Dance is one of six books by Henry
Brewis published by Farming Press.
The others are:

Chewing the Cud
The farming scene from a peasant's-eye view –
the third cartoon collection.

Clarts and Calamities
The diary of a year in a peasant's life with its
disasters (frequent) and triumphs (rare).

Don't Laugh Till He's Out of Sight
Stories, verses and illustrations revealing the
hazards awaiting anyone venturing on life as a
farmer.

Funnywayt'mekalivin'
The first collection of Henry Brewis's cartoons
featuring Sep, the universal peasant.

The Magic Peasant
The second cartoon collection showing Sep in
his world of collie dogs, auctioneers, sheep and
the long-suffering wife.

For a free complete catalogue please contact:
**Farming Press Books,
Wharfedale Road, Ipswich IP1 4LG.**